THE SCIENCE OF
MINECRAFT

THE REAL SCIENCE BEHIND THE CRAFTING, MINING, BIOMES, AND MORE!

JAMES DALEY

Skyhorse Publishing

TABLE OF CONTENTS

PROLOGUE

A few minutes before my life was forever changed, I was just an ordinary young scientist, dutifully working at the Mojang Institute of Novel Experimentation as a Certified Research Associate (First Tier), cleaning up my lab after a long day of experimentation. Like most days, I had remained at work far past when all my colleagues had gone home for the evening, as I greatly enjoyed the quiet solitude of the empty campus after dark. The last thing I needed to do before I could go home was put my last three test tubes into the cold storage container, but I had run out of test tube stoppers and had to get some more from the storage closet in the hallway.

When I got there, I found only an empty box, meaning I would have to venture all the way down to the warehouse in the basement to find another. I sighed, judgmentally shaking my head at the carelessness of my colleagues, and began to make my way down the winding hallways and many staircases that led to the basement.

Since I had just begun working at the Institute a few weeks prior, I'd only been down to the warehouse once before, and that was with my boss leading the way. Still, it never occurred to me that I would get lost on my way down to grab a new box of test tube stoppers—but that's exactly what happened.

To be honest, I'm not sure if I was ever going in the right direction, because before I even made it to the basement stairwell, I found myself all turned around and in a part of the building I had never been to before. I paused for a second to look around and get my bearings, wishing I had done this before everyone went home for the night, when a pair of double doors at the end of the corridor burst open, and a frazzled-looking woman in a white lab coat came running out, looking quite out of breath and seriously freaked out.

Her eyes went wide when she saw me. "You!" she yelled. "Stop right there!"

My first thought was that I had ventured into one of the restricted areas of the campus and that I was about to get a serious scolding. "I'm sorry," I said. "I'm just trying to find my way to the warehouse for a box of—"

But before I could finish my sentence, the crazed woman had cleared the distance between us and cut me off. "What's your job here?" she asked, her voice shaking.

"Uh . . . Certified Research Associate."

"What Tier?"

"First," I said, hoping this might get me out of trouble.

"It'll have to do," she said, grabbing my hand. "Come with me."

Before I knew it, she was leading me back through the doors she had just burst open and down a series of hallways until we came to a very heavy-looking metal door with a sign on it that read CLASSIFIED CLEARANCE ONLY.

The woman retrieved an access card from her pocket and I noticed that it identified her as "Dr. River Song, PhD – Chief Scientist, Department of Obscure Experiments." She swiped her access card on the electronic lock and the door swung open, revealing a tableau that quite literally took my breath away.

It was a laboratory, not unlike my own, but right in the center of it was the strangest thing I had ever seen. It was a huge gray frame as tall as the ceiling and seemingly made of large metallic blocks that glistened as if they were coated with a thin sheen of crystal. Inside the frame was a glowing green substance that I could not identify, but almost seemed to be made from some kind of oily liquid.

I was about to ask what this strange object was when I saw the man lying on the floor just in front of the frame, his eyes barely open, breathing so shallowly I was certain that he did not have very many breaths left in him.

I instantly recognized the man, though I could not imagine what he was doing there. It was none other than the legendary physicist Charles

Benzak, PhD—one of the greatest scientists to have ever worked at the Mojang Institute of Novel Experimentation.

"Is that Dr. Benzak?" I whispered to the frazzled Dr. Song. "I thought he retired years ago."

"Yes, it is," she replied. "Though he never retired." She pointed at the strange object in the center of the room. "He's been on the other side of that."

Now I was really intrigued. "But what is that?" I asked.

Dr. Song shrugged. "I don't know exactly. Some kind of doorway. To someplace . . . different."

"Different how?"

She looked me square in the eyes. "That's what I need you to find out."

A chill ran up my spine. "You want *me* to go through that thing?"

"Yes," she replied, the urgency in her voice growing. "It will only stay open for another minute, and I have to see to Dr. Benzak."

"And what do you want me to do, exactly?"

"Research," she replied. "Experiment. Document everything you find that's different from Earth."

My eyes went wide as silver dollars. "What do you mean by 'different from Earth'?"

"There's no time to explain," she said, now pushing me toward the strange object. "Your contract requires you to accept any assignment we give you, and this is your new assignment."

Dr. Song was right, of course. I had agreed to those terms when I took the job. What could I do? I began to walk toward the strange object, growing more and more nervous with every step. Right before I was about to step through, Dr. Benzak opened his eyes and looked right at me, a hint of sadness on his face.

"You got any advice for me?" I asked him.

He struggled to take in enough breath to form words. Finally, his lips parted. "Mine," he said, his voice barely a whisper. "Craft."

Then the legendary Dr. Charles Benzak closed his eyes again, and the woman who had brought me to that strange laboratory shoved me through the portal.

CHAPTER 1

LEARNING TO CRAFT

I opened my eyes to find myself in a forest, though it wasn't like any
forest I had ever seen before. At first, I was a bit too disoriented to
notice the difference, glaring as it was. There were great, towering trees
rising out of a grass-covered hillside. There were rocks and shrubs and
a blue sky above, and a brook babbling somewhere off in the distance.
Something was just . . . different.

Maybe it's the grass, I thought. Every inch of this forest seemed to be
covered with tightly clipped grass as far as the eye could see. I wondered
who could possibly maintain such an enormous lawn, and why they would
bother to do so.

But no, I thought. That wasn't the strange part.

Was it the trees? They were all incredibly straight, super tall, and not
one of them had a single branch extending from a solitary trunk.

No, that wasn't it either.

I took off my glasses, rubbed my eyes, and put them back on . . . and
then I saw it.

Squares. Everywhere I looked, everything was square—the trees, the
shrubs, the rocks, *even the clouds*. All squares. Or rather, I realized as
I approached a small boulder, not squares exactly—*cubes*. Everything
everywhere was made of identically sized cubes.

Well, almost everything. There was an animal that appeared to be a
sheep standing on the hillside, staring at me warily, who seemed to be
made of slightly smaller cubes than the rest of the environment. And much
of the smaller vegetation was actually comprised of intricate matrices of
imperceptibly thin squares.

1

Of course, my first thought was that I must be hallucinating or dreaming or something (and I admit that I held this as a possibility for quite some time), but before I got a chance to think much about this, the sun began to set, and I was enveloped in darkness.

That's when I noticed the moon—the perfectly square moon—moving at breakneck speed across the star-studded sky.

For a second, I allowed myself to almost feel excited by the sheer scientific hugeness of what I was experiencing, but it didn't last. You see, a second or two after I noticed the square moon, I noticed that it had illuminated a humanlike figure approaching me from the shadows.

"Hello!" I called out to the approaching figure.

There was no reply. I was about to call out again, but the mysterious figure emerged into the fullness of the moonlight, and I was instantly gripped by fear.

At the time, of course, I had no idea what this creature truly was, but I did know what it looked like: a human-sized, square-headed, black-eyed, green-faced zombie.

So, what else could I do? I turned and I ran.

Within a few seconds, however, I turned around a tree and ran straight into another zombie. Startled, I began to wildly swing my fists at the beast, hoping at least to scare it away. I landed a few punches on its head, and then (I swear this is true) the whole thing just disappeared in a puff of smoke.

For a moment, I was too stunned to move. Did I do that? Where had it gone? How was that even possible?

I didn't get to hypothesize any answers to my questions, though, because right at that moment another zombielike creature emerged from behind a tree—this one even more grotesque than the last. It had no arms at all and tiny little legs and a horrified expression on its blotchy green face. I punched this one too as I darted around it, but instead of disappearing it began to hiss and smoke. When I turned to see if it was following me the whole creature just exploded with incredible force, creating a crater in the dirt beneath the spot where it had been standing.

That's when I really started to run.

I darted through the forest as another two terrifying creatures appeared out of the dark: an enormous spider with glowing red eyes and a skeleton with a bow shooting arrows wildly in my direction.

I turned and sprinted in the opposite direction of where the beast had appeared in front of me, running and running until I found myself up against a dirt hillside that was far too steep to climb. Turning around, I saw that the creatures were still pursuing me, and they seemed to have brought friends.

Desperately, I began trying to claw my way up the hillside. That's when I discovered one of the other great mysteries of this world.

You see, when I grabbed hold of the hillside to pull myself up, the hillside began to disappear beneath my hands, one solid cubic meter (35 cubic feet) at a time.

Later, I would learn that these chunks of dirt did not actually disappear at all, but in fact simply shrunk down to a fraction of their original size and, for some reason, appeared in one of my many pockets (I had thirty-seven pockets, I later figured out—seriously, thirty-seven), but at that moment I was far too scared to realize any of this. I just dug and dug into the dirt hill until I came out the other side and could keep on running.

And then before I knew it, the sun was rising.

How long was that night? I wondered. *Fifteen minutes? Ten minutes?*

In any case, the creatures seemed to have stopped following me by then and I wanted to put as much distance between myself and them before night came again . . . which, based on what I had seen so far, might be just a few minutes away.

As I ran through the forest, I began to notice subtle changes in the environment. The tall trees with dark brown bark began to give way to smaller trees with lighter brown bark, and then other small trees with black-speckled white bark (almost like a birch tree, I thought, if birch trees were made entirely of right angles and grew with meter-wide [foot-wide] trunks). I ran past some more cubic sheep, but also some cubic pigs and cows and chickens, and even a cubic wolf that tried to bite me.

Finally, as the sun began to set again, I found myself in a swampy part of the forest where the ground was spotted in shallow, murky ponds, and long vines hung down from every tree. Not wanting to be stuck in a swamp when darkness came (if the idyllic forest spawned giant spiders and zombies, I didn't even want to know what the scary swamp would spawn!), I picked up my pace, just hoping to find some kind of shelter before dark.

I came to a body of water that seemed a bit too deep to wade through, and I was just about to try a different direction when I spotted what appeared to be a small stone house on the other side. So, I jumped into the water and swam my heart out, reaching the little house just as the sun dipped beneath the horizon.

Unfortunately, when I finally stumbled up to the structure, I immediately realized that the house had no door, and the spaces where the windows used to be were just empty cubes of space and cobwebs. Still, it seemed better than being out in the forest, so I ran inside anyway and crouched in the corner to hide.

That's when I discovered all those blocks of dirt in my pocket. They were much smaller than the blocks I had pulverized while I was digging that tunnel through the hillside, but there were loads of them in there. I took one out and held it up in the fading light, wondering what laws of physics could possibly make such a thing possible. I didn't allow myself the luxury of contemplating this mystery, though, as I quickly realized that there just may be enough of these tiny blocks of dirt to stack in the doorway and create some type of barrier against a monster.

To my great surprise, however, the moment I placed one of these tiny dirt blocks down on the threshold of the doorway, it expanded all the way back to its original dimensions, instantly filling the bottom half of the empty space.

Okay, I thought. *That's . . . interesting. I guess I should try to do that again.*

I took another tiny block of dirt and placed it on top of the first one. Just like before, it instantly expanded back to its full dimensions, filling the rest of the doorway without the slightest gap.

My mind raced with the scientific implications of this discovery, but I quickly quieted my curiosity and set to work filling in the windows with dirt blocks, as well as filling a hole in the roof that I hadn't noticed before.

Just when I finally got myself fully enclosed (and completely in the dark, by the way) I began to hear the moans and footsteps of those horrible creatures just beyond the walls.

All night they kept at it, and it wasn't until the sun had fully lit the sky that I summoned the courage to break down my dirt door and venture back outside.

I found that my little sanctuary of a shack was, in fact, just the outermost building of what appeared to be a long-abandoned village of some kind.

That's when I thought back to Dr. Benzak's mysterious two-word instructions: *Mine. Craft.*

Maybe the answer to what he meant can be found somewhere in this village, I thought.

So, I began to explore, first just taking in the village as a whole and then looking through each of the many decrepit buildings one by one. The first few structures I searched contained either nothing but cobwebs, or else a solitary bed. Finally, though, I came upon a small stone house that, in addition to a bed and chair, contained a small wooden chest. Inside the chest I found some blocks of wood the exact same size as the miniaturized dirt blocks and an apple that seemed (despite having been there for what I imagined was quite a long time) to still be perfectly fresh . . . and I was starting to get rather hungry. I ate the apple, which tasted more or less like any apple I had ever tried on Earth, and then went out to explore some more buildings.

That's when I found the object that finally offered the first clue about Dr. Benzak's instructions. Like everything else, it was a cube about a meter (3 feet) squared, and it appeared to be made of some wood-like material. There were some knobs and dials on the outside and a drawing of a 3x3 grid of squares on the lid. Clearly, this was nothing like I had ever seen before, because when I touched it, the top opened to reveal a graphic interface with pictures of various items on the left, empty squares on the right, and the word "Crafting" at the top.

I felt a chill go up my spine. This must be what Dr. Benzak meant by "Craft." But how do I use it?

There were thirty-five squares on the left side, each containing a different item. I noticed right away, however, that most of the squares were red and the object pictured in them slightly faded, while only five squares were clear.

I touched each of the squares in turn to see what they were: stick, bowl, oak button, oak pressure plate, and crafting table.

I closed the lid and took a step back. Just as I suspected, the "crafting table" pictured in the device looked exactly like the device itself. Opening the device up again, I noticed that touching the picture of the crafting table made four wood blocks appear in the grid on the right. Seeing as I happened to have exactly four blocks of wood, I placed them in the grid on top of the squares, and then closed the top.

The box made some whirring and clicking noises, and then out popped a miniature version of the crafting table itself.

At first, I was disappointed by the size of the item (wondering if I had just wasted those four blocks of wood on a model of some kind), until I remembered how the dirt blocks had expanded back to their original size when I placed them down. So I picked up the mini crafting table, put it in my pocket, and opened the original one up again to see if there was anything else it could make. But when I looked at the graphic interface again, there were no pictures on the left side at all.

Interesting, I thought. *It must only show the objects that I actually have the materials to craft. Amazing!*

I quickly started exploring the rest of the buildings in this abandoned village to see if I could find any other useful materials. All in all, I wound up finding some sticks, a few stones, a loaf of bread, and a pair of leather pants. I took everything I found (even the leather pants, which were never much my style back on Earth), and then barricaded myself back in the shack with the crafting table before night fell again. Once again, the village was overrun by monsters after the sun went down, but this time I was not quite so lucky to make it through the night unscathed.

Just before sunrise, I heard the now-familiar hissing of the exploding zombie monster just outside my shack, and before I knew it—*BOOM!*

I was thrown to the ground, and when I finally got back on my feet, I saw that one whole wall of the structure had just evaporated, along with the chest and crafting table that had only seconds ago been sitting on the floor beside it. Clearly, the explosion had impacted me too, as my whole body began to throb with pain. Thankfully, I still had my miniature crafting table and the items I had found in my pocket, so I took a deep breath and began to sprint out toward the edge of the village.

I ran through a veritable army of zombies, spiders, and other monsters before finally making it back to the edge of the forest. I kept running until the sun, at long last, had risen over the horizon, and I could finally stop to catch my breath.

At this point, I found myself in a forest that seemed to be made up entirely of birch trees. I decided to look for a suitable place to set up some sort of camp, when one of those giant spiders hopped out from behind a tree, practically right in my face.

Without thinking, I just closed my eyes and started punching wildly. I'm not sure how many punches it took to kill the spider, but at some point, I noticed that the spider was gone, and I had actually been punching the tree that was behind it. Just as I realized this, the section of tree that I had been punching popped right out. Like the dirt from the hillside, it shrunk down to a fraction of its original size, leaving a sizable gap right in the middle of the trunk.

Yup, I had literally just punched a hole in the middle of a birch tree, which of course made me recoil back in terror, assuming that the top portion of the tree would now come falling down on top of me. But I discovered another way this world was very different from the one that I had come from.

To my complete amazement, the tree did not fall. It didn't even wobble. I touched the top portion, and it felt just as solid as the bottom portion. I punched through the tree stump, which popped off as easily as the last piece, and then stared in awe at a tree trunk that began a solid two meters (six and a half feet) above the ground, hanging there in space as if it were completely unaffected by gravity at all.

I decided that this was probably something I should investigate in more detail later, but right now I needed to get back on track to survive the next night. I kept on punching away sections of trees and collecting the wood from them, figuring that if I collected enough I could build a more suitable shelter for myself someplace less conspicuous than the middle of an abandoned village.

I figured my best bet would be to build my new shelter at the top of a hill where I could easily see any dangers as they approached, and hopefully gain some protection from the height as well. It took me all day and most of the logs I had collected, but by nightfall, I had a simple, square wooden hut that I sealed myself inside of, and I was once again encased in darkness.

Although I did hear some monsters outside during the night, they seemed to be far fewer in number than the hordes that had been roaming through the village. After a mostly uneventful night, I finally felt as if I had some basic level of safety.

So, feeling optimistic and accomplished, the next morning I set up my crafting table in the shack and began my work of discovering what it could do.

***

By the time the sun rose on the first day of my research, I was no longer giving a thought to the strange twist of fate that had landed me on Planet Minecraft. After all, I had taken the job at the Mojang Institute because I wanted to discover things that no one else had ever discovered before, and if I couldn't do that here, then I couldn't do it anywhere.

Clearly, I realized, the best place for me to start was with the crafting table. Now, as I had discovered the day before, the items the crafting table would let me make were limited by the materials I possessed when I opened it up, and at the moment, all I had was three blocks of dirt, three blocks of stone, and six logs. I didn't think the table would give me too many options; as it turned out, I was right. I could only craft three items with the materials in my possession: oak wood planks, a cobblestone slab, or a block of solid oak wood.

Figuring I couldn't go wrong with so few options, I chose to make some oak wood planks. According to the crafting recipe, this only required a single oak log, so I took one of the logs (still in miniature form, obviously) out of my pocket and placed it in the grid at the top of the crafting table. I closed the lid, heard a bit of strange whirring and knocking sounds come out of the table, and then the lid popped back open to reveal four small cubes of oak wood planks.

Wow, I thought to myself. *One log makes four blocks of planks. That's awfully efficient.*

As soon as I picked up my new oak planks (which I had no idea what to do with) I was amazed to see a whole new selection of items pop up on the screen, including another crafting table, sticks, and a few other wooden items that looked to be mostly decorative. Considering I had nothing to put in a bowl and I already had a crafting table, I went with the sticks.

This is when things started to get interesting.

As soon as I removed the sticks from the table, even more items became available to craft, the most interesting of which were a stone sword, a stone ax, and a stone pickaxe. My first inclination was to make either the sword or the ax, as these two seemed like they would have been the most helpful on the previous day, considering I spent so much time fighting zombies and punching holes in trees.

But then I remembered that "craft" was not the first word Dr. Benzak had spoken. The first word was "mine."

Of course! I thought. *I can use the pickaxe to mine!*

You see, what I realized right away was that this table only let me create things if I already had the materials to make them, which means that if I was going to really see what the crafting table could do, I would have to get more materials. Clearly, what Dr. Benzak was telling me was that the best way to get materials was to mine for them, and for that, I would need a pickaxe.

After crafting the stone pickaxe, I was quite happy to see that I still had enough materials for either a wooden sword or a wooden ax (though not, I realized, for both). Figuring that I had done a better job punching trees than monsters, I decided to go with the sword.

Thus, with my stone pickaxe and the wooden sword in hand, I went out to find a good spot to start my first mine.

Leaving my little hut for the first time in daylight, I realized that the hill I had chosen to build it on was right on the edge of two different biomes: the birch forest I had seen the day before and a sort of red rock, desertlike biome that looked a lot like Badlands National Park in South Dakota back on Earth. Thinking it might be easier to start a mine someplace where the rocks were a bit more exposed than in the forest, the first thing I did was venture into the Badlands biome.

Even before I began to dig my first mine, I realized that this would be a far safer place to make my home than a little shack on a tiny hill in the middle of a forest. In such a barren landscape, I would be able to see the zombies long before they reached me, and the red clay mesas offered the extra protection of their considerable height, as well. After spending some time searching for a suitable location, I decided that the safest thing to do would be to carve my new home into the side of one of the red clay cliffs. I had read about the Native Americans doing that in this kind of landscape back on Earth. So I decided to forego digging my first mine, and instead, I went straight back to my shack, punching a few trees along the way so I could make myself a new crafting table to take with me.

***

After a short time, I had managed to carve out a good-sized cliff dwelling in a red clay cliff just inside the Badlands biome. The digging itself, which I had assumed would be the most taxing of all the tasks I had to do, wound up going far more smoothly than I imagined it would. One of the key facets of this strange world in which I found myself is that digging with a simple pickaxe, even in great quantities, is extraordinarily efficient. Furthermore, one of the most difficult parts of such a task on Earth would be collecting and transporting all the clay I was digging up away from the dwelling, but this was easily accomplished simply by picking up all of the miniaturized blocks and putting them in my pockets until all thirty-seven of my pockets were full. Considering that each of my pockets could apparently store sixty-four blocks of clay, this meant that I only had to make two trips back down to the bottom of the mesa before I had carved out a sizable home for myself.

I carved my dwelling ten meters (33 feet) above the bottom of the cliff and five meters (16 feet) below the top of it, with a five- by ten-meter (16- by 33-foot) opening that looked north toward the birch forest and whatever was beyond that. I had a good deal of confidence in my dwelling's ability to keep me safe from most of the monsters, given that I had not yet seen any that had the ability to fly. That said, I was a little bit concerned about the spiders. If they were anything like Earth spiders, climbing up this wall wouldn't prove much of a challenge to them, so I would have to stay alert when night fell, at least until I got a sense of their ability to make it up the face of the cliff. Worst-case scenario, I figured I would have to fill the opening of my cliff dwelling at night, and then just carve it back out in the morning.

With my cliff dwelling completed, the only thing to do was wait for nightfall and see how it held up. I still hadn't figured out how to make a bed yet, but I thought that was probably for the best, as I wanted to remain as alert as possible through the night.

As soon as it got dark, I began to see spiders crawling around at the base of the cliff beneath my dwelling. At first it was just two of them, but more and more started to appear, crawling out of the holes and caves that dotted the landscape. After a couple of minutes, there were perhaps a dozen, or maybe more. Terrifying creatures, the spiders had glowing red eyes and seemed to be the size of large Rottweilers. There were also a few other types of creepy monsters out there as well, including some of those green zombie guys, and quite a few of the exploding variety as well. As the night wore on, some of the monsters ventured closer and closer to the bottom of the cliff, but thankfully none of them seemed capable of climbing it. At least not yet.

All in all, I considered my first night in my new home to be a resounding success. While all the creatures that plagued me on previous nights were clearly visible all around the base of the mesa, none of them had gotten close to making it inside. Having successfully made it through my first night without being assaulted by monsters, I now felt free to spend some time mining, crafting, exploring, and researching everything this new world had to offer.

* * *

What follows in the subsequent chapters is a curated selection of the highlights of my research. Although I could easily have filled a dozen books of this length discussing all of the interesting and amazing things that I discovered on Planet Minecraft, I tried to stick to the mission I was given and include only the facets of this world that are particularly different from Earth, and which I was able, if not to explain, at least to come up with some plausible scientific theories about. Of course, I still do not know if anyone back on Earth will ever get to read any of this, but that's a matter for another day.

CHAPTER 2

MINECRAFT GEOLOGY

Without a doubt, one of the most fascinating aspects of scientific exploration here on Planet Minecraft is exploring the geology of this strange new world. Clearly, whoever created this crafting table technology thought so as well, as it is quite easy to use the device to determine the local names of different rocks and minerals, thus allowing me to compare them to their terrestrial counterparts. What follows is the result of many expeditions all over the terrain of this new world, mining deep into the ground, climbing high up the mountains, and nearly getting myself killed in a hundred different ways. Please note that I am purposely not including a description of every rock, mineral, and geologic formation I studied here, as I have neither the time nor the inclination to do so, and many of them are, well, just not all that interesting.

LAVA, MAGMA, AND VOLCANOES

One of the first things I noticed when I began exploring Planet Minecraft was the incredible amount of volcanic activity on this planet. Seriously, there is lava flowing out of the ground practically everywhere you look. You find volcanoes near the surface of caves, flowing out of the side of cliffs and mountains . . . you can pretty much pick any spot on the surface of this planet, and if you dig down deep enough, you're going to find some lava. As a scientist, I found this quite exciting, because on Earth, the existence of volcanic activity tells you that you will be able to see many fascinating geologic processes at work right in front of your eyes. It doesn't quite work that way here on Planet Minecraft, though. The lava here is . . . well, it's just kind of crazy. I actually don't even know where to begin with it. In some ways, it's a lot like the lava back on Earth, but in so many ways it's very, very different.

Before I get into the deep weirdness of Minecraft volcanism, let's take a moment to talk about how volcanoes function back on Earth.

Now, if you could take a big old knife and cut the Earth in half like an onion, the first thing you would likely notice is, once you get down past a few relatively thin layers near the surface, pretty much everything else is a turbulent, hot, glowing ball of magma. Magma, of course, is basically just rock (usually basalt) that has reached a high enough temperature to melt and remain in a liquid or semi-liquid state under the surface of the Earth.

It is important to note, before I get any further into this topic, that the terms "magma" and "lava" have a very specific distinction back on our planet that does not seem to be the case here on Planet Minecraft—at least, according to the labels in my crafting table. On Earth, this distinction is simple: hot, molten rock under the ground is called magma; when that hot, molten rock comes up out of the ground in liquid form, it's called lava (at least until it cools and solidifies, at which time we just call it rock, obviously). Here on Planet Minecraft, the distinction is also quite simple; it just happens to be completely different. Here, all of the hot, molten rock is called lava whether it's above ground or below ground, while magma is a solid, slightly less hot version of lava that also keeps the same name whether it's above or below the ground.

Anyway, back to Earth. The surface of Earth rests on a layer of solid tectonic plates that float on top of this underground sea of magma. These plates are in constant motion, though the motion is so slow that we can almost never actually feel it from the surface. The only time we can feel these plates move is during an earthquake, which is essentially just two of these plates rubbing up against each other with enough friction to send vibrations emanating up to the surface.

Now the important thing to keep in mind is that Earth's tectonic plates do not form a solid, smooth layer between the magma beneath and the surface above. Not only are there gaps where the plates overlap each other, but there are also fissures that can occur even at the center of very large tectonic plates. Since the magma is under enormous pressure, it is constantly trying to rise up and escape through any gap or fissure large enough for it to pass through. Often, these gaps or fissures don't reach all the way up to the surface, and the magma simply fills the empty spaces, forming large magma chambers in the bedrock. When we see magma emerging from the earth as lava, it's because the heat of the magma has created so much pressure inside the chamber that it bursts out through the surface, which can happen very quickly as a powerful explosive eruption, or quite slowly as a dribbling effusive eruption. In either case, the eruption will continue until the pressure in the magma chamber decreases enough that it is no longer able to force the molten rock all the way up to the surface.

Okay, so that's how volcanoes work on Earth, but what about here on Planet Minecraft?

Well, first let's talk about the similarities. For one thing, lava on Earth is an extremely viscous liquid, and as such, flows considerably more slowly than less viscous liquids like, for example, water. Here on Planet Minecraft, the same seemed to be the case when I first began observing the behavior of lava, but I wanted to be certain, so I conducted a simple experiment to test it out. For my experiment, I went to the ravine near my home where I knew there was both lava and water flowing out of a sheer rock cliff. Starting with the water, I built a slope (well, more of a staircase really) exactly ten meters (33 feet) out and ten meters (33 feet) down from the spot where the water was coming through the rock. With my staircase constructed, I blocked the source of the water with a cube of granite and waited for the water that had already emerged from the rock to stop flowing. Once the water stopped, I simply removed the block of granite and timed how long it took the water to reach the bottom of the staircase.

The result: 5.88 seconds.

I then repeated the same exact experiment with the lava and came up with a whopping 35.2 seconds—clearly far slower than water.

So that's one similarity: lava is super thick on both planets. Are there other similarities? Well, there is the fact that the lava on Planet Minecraft is very hot, just like the lava on Earth is very hot. How hot exactly? Unfortunately, there isn't a recipe in the crafting table for a thermometer, so I don't really have any way to determine that precisely, but it certainly is hot enough to burn pretty much everything I dropped into it, even a diamond I traded for with a villager (at great cost, by the way). Still, though, to be scientific, I figured the best way to determine if the lava here was roughly the same temperature as the lava on Earth would be to see if it would melt both quartz and iron. You see, lava on Earth ranges from about 700 degrees Celsius (1292 degrees Fahrenheit) to 1250 degrees Celsius (2282 degrees Fahrenheit) at its hottest. At the same time, quartz has a melting point of about 600 degrees Celsius (1112 degrees Fahrenheit), while iron has a melting point of about 1538 degrees Celsius (2800 degrees Fahrenheit). So, to determine if the lava here on Planet Minecraft is at least in the same temperature range of the lava on Earth, I simply needed to find a pool of lava, throw in some quartz and some iron, and observe the results. If the lava can't melt either one, then it is cooler than the lava on Earth; if it can melt both, it is hotter; if it can only melt the quartz, it is at least roughly in the same temperature range as terrestrial lava.

Here are the results of this experiment: I tossed both a block of quartz and a block of iron directly into a pool of lava and watched as they both evaporated more or less instantaneously. Obviously, I expected the quartz to burn right up, but I was honestly not expecting the iron to burn as well (or at least not so fast). What this tells me is that the lava here is at least 1538 degrees Celsius (2800 degrees Fahrenheit), if not considerably hotter.

Here's where everything gets a bit wacky, though. First of all, the lava here seems to be kind of picky about when it decides to burn things and when it doesn't. For example, throwing a block of quartz into lava makes it instantly melt, burn up, disintegrate, or something (I don't know, I could never find any remnants of it afterward), just as you would expect. However, when I built a nice little bathtub out of quartz blocks at the bottom of a lava flow, the bathtub filled right up with the molten rock and stayed completely intact. No charring, singeing, melting, or anything. How, you might ask, is it possible that the same material that burns up instantly when thrown into a pool of lava is also capable of easily containing it? I have no idea. I'm not even going to try to come up with a theory for it . . . at least not yet.

Another frankly mystifying aspect of the lava on Planet Minecraft is its density. You see, on Earth, lava is extraordinarily dense (I mean, it's literally made out of melted rocks, after all), which means that basically everything you throw into it should float, at least until it burns up or melts. But that's not what happens. On the contrary, everything that I ever threw into a pool of lava sunk almost instantly. This is especially strange considering how slowly the lava flowed down the staircase in my earlier experiment.

GLOWSTONE

Certainly one of the more interesting minerals here on Planet Minecraft, glowstone seems to occur naturally through the portal into the dark cave "dimension" (or whatever that place is) of the Nether. This luminescent mineral occurs in large blobs that hang from the ceiling of the Nether cave, and when struck with a pickaxe it breaks down instantly into dust, which can easily be transformed back into full-sized, one-meter-square (3-foot-square) glowstone blocks simply by combining four piles of glowstone dust in the crafting table. With an appearance not unlike this world's version of magma, glowstone in its natural state appears dark amber in color with large patches of glowing orange spotting its entire surface.

What makes glowstone so interesting is the incredible amount of light it puts out—every bit as much as my best lantern—without creating any discernible heat whatsoever. The usefulness of such a material is hard to overstate. Not only is it a seemingly infinite source of illumination (which, I admit, is not nearly as strange in this world as it is in our own), but it manages to accomplish this while also maintaining enough structural integrity to be useful in building a wide variety of weight-bearing structures. If you are actually capable and brave enough to collect a large quantity of glowstone from the deep reaches of the Nether, you could easily build an entire house or even a castle out of the magnificent mineral (though I wouldn't use it for the exterior walls myself, as it offers little to no resistance against the exploding creepers).

Needless to say, we have nothing quite like this mineral on our planet. In appearance, the closest thing I have ever seen on Earth is pallasite, which is a very rare meteorite composed of golden-green crystals suspended in a metallic web of iron and nickel. While a block of pallasite one meter cubed (35 cubic feet) being held up in the sunshine would indeed look quite a bit like glowstone, it does not, alas, produce any light of its own. The closest thing we have on Earth to this type of luminescent mineral would probably be phosphorus. When exposed to oxygen, phosphorus will emit a faint glow through a process called chemiluminescence; of course, this light is nowhere near as bright as the light emitted by glowstone.

OBSIDIAN

Obsidian here on Planet Minecraft is quite an interesting substance. Now, most of the rocks and minerals here that have the same names as their counterparts on Earth (at least according to the crafting table) have wound up being fairly similar in appearance, usage, and most other qualities. Minecraft obsidian, on the other hand, is quite different than Earth obsidian. While the two do look quite similar from a distance—both nearly black and quite reflective—when inspected up close the Minecraft obsidian has more of a purple tone than the Earth obsidian, which tends to be almost greenish in hue. Both Earth and Planet Minecraft obsidians are formed through a notably similar geologic process, but that is really where the similarities end.

On Earth, obsidian isn't actually a rock or a mineral at all. Rather, Earth obsidian is a volcanic glass, formed when highly viscous, lightweight lava reaches the surface of the Earth and cools too quickly for any actual crystals to form. This lack of crystals is what makes it a glass as opposed to a mineral. Being a glass, obsidian is extremely fragile on Earth, shattering quite easily and without a great deal of force needing to be applied to it. This is perhaps the greatest difference between the obsidian on Earth and the obsidian on Planet Minecraft. On this planet, obsidian is actually one of the hardest and most resilient materials I have found anywhere. It takes an enormous amount of effort to mine it, and only the hardest and strongest of pickaxes can be used to break it down at all. Additionally, it is extraordinarily resistant to explosions from the exploding zombies, as well as from any TNT-based explosions I have managed to subject it to in my experiments.

Now, as I mentioned above, the obsidian here on Planet Minecraft does seem to form in a manner very similar to that of the obsidian on Earth, which is to say, nearly instantaneously by rapidly cooling lava with water.

On Earth, it is precisely this rapid cooling that causes obsidian to have its signature lack of crystalline internal structures (which, of course, is precisely what makes it so brittle and prone to shattering). Fascinatingly, even though both obsidians are formed through such clearly similar processes, the resulting materials have almost completely opposite properties in virtually all areas except appearance.

It is also worth noting that there are some properties of obsidian here that simply go beyond my current understanding, as it is the key material needed for creating the interdimensional teleportation device that forms the doorway to the Nether (but more on this in a later chapter).

NETHERITE

Perhaps the rarest of all materials here on Planet Minecraft, Netherite, as its name suggests, can only be found on the other side of the Nether portal (and even there, it is extremely difficult to locate). For one thing, it does not form in full-sized, one-meter-square (3-foot-square) blocks, the way that nearly everything else here on Planet Minecraft does. Rather, it must be smelted down from a seemingly fossilized substance known as Ancient Debris, an extremely hard and resilient block that almost has the appearance of a petrified log on Earth.

Actually finding enough blocks of Ancient Debris to create Netherite is in itself a monumental task. Even on my longest expeditions into the Nether, when I would bring all manner of mining equipment and TNT with me and reduce vast sections of the Nether bedrock to rubble, I would only find the smallest deposits of Ancient Debris, which then had to be mined with a diamond pickaxe and an extraordinary amount of effort (as no amount of TNT seems to harm it whatsoever).

Once I had managed to create some actual Netherite, however, it proved quite easy to craft into weapons, tools, and armor, the strength of which has no equal either in the Minecraft world or on Earth. Decidedly stronger than even diamonds, the only possible equivalent in our own universe would probably be lonsdaleite (which is almost as rare in our universe as Netherite is in this one). Lonsdaleite is a diamond-like crystal that is only known to form naturally when graphite-rich meteors strike the Earth. The enormous heat and pressure created by the resulting explosion of this meteor strike transforms the graphite into a mineral very similar to diamonds, but with a hexagonal crystal structure instead of a cubic crystal structure. In theory, if one could find enough of this mineral in its pure crystalline form, it would be nearly 60 percent harder than diamonds; however, no one on Earth has ever been able to find (or even synthesize, for that matter) lonsdaleite pure enough to actually exhibit these properties.

Another interesting fact about Netherite is that it is a key material in the crafting of a lodestone: the highly magnetic block capable of altering the direction Minecraft compasses point even from great distances. While I was not able to observe any magnetic properties exhibited by Netherite on its own, for some reason when combined with ordinary stone in the crafting table, this extraordinarily strong magnetic effect is produced.

DIAMONDS

While there are quite a number of things that I often daydream about bringing back to Earth if I ever get out of this place—redstone dust, Netherite, one of those fully functioning boats that shrink down to the size of a laptop—I will certainly fill at least one of my pockets with a whole bunch of Minecraft diamonds. Much like their Earth counterparts, diamonds here are relatively rare, and take quite a bit of mining to locate. That said, the ease of mining in this world means that I have frequently been able to spend a leisurely afternoon digging around some of the deeper mines I have created and come back with a dozen diamonds.

Much like on Earth, diamonds here on Planet Minecraft have both great aesthetic and utilitarian value. They are sparkling, reflective gemstones with exceptional hardness that makes them ideal for use in cutting nearly any other material. That's about where the similarity ends, though.

First, let's talk about the color. The diamonds here on Planet Minecraft are all the same color: light bluish green. On Earth, diamonds can certainly be blue, and they can certainly be green, though I have never seen any with quite this particular shade. That is in part because Earth diamonds wind up either blue or green based on very different circumstances in their formation. Blue diamonds are formed when trace amounts of boron make their way into the diamond while it is crystallizing, while green diamonds are formed when diamonds are exposed to radiation, usually as a result of being in the vicinity of naturally occurring uranium. Again, while some natural blue, boron-based diamonds can have a bit of a greenish hue, and some radiation-affected green diamonds can have a bit of a bluish hue, I've never seen the bright aquamarine of these diamonds occur anywhere naturally on Earth (although, of course, some lab-grown diamonds may have a very similar appearance, but none that occur naturally).

Of course, diamonds on Earth are famous for their extreme hardness, making them ideal for industrial uses such as cutting through rock or making fine surgical instruments. The diamonds here are no exception—they are also quite hard and make excellent cutting tools— but the Minecraft diamonds are hard in a way that no diamond on earth could ever be: they are hard in every possible direction.

Allow me to explain. You see, like all crystals, Earth diamonds gain their strength by the symmetrical arrangement of their atoms. In diamonds, you have tightly bound carbon atoms in a diamond cubic structure that makes them nearly indestructible, but only from certain angles. Like all crystals as well, diamonds have what are known as cleavage planes, which means that there are angles you can strike a diamond perpendicular to the strongest arrangement of its atoms, in which it will not be able to withstand very much pressure at all. This is why, on Earth, you would be able to make a very strong drill bit out of diamonds, but if you tried to make a window out of diamond it would shatter almost as easily as ordinary glass.

The diamonds here on Planet Minecraft, however, somehow manage to maintain their extraordinary strength from all directions. This has some truly magnificent applications, probably the most useful of which that I have found has been in the construction of armor to protect myself from the many dangerous life-forms here on this world. A diamond armor chest plate made out of Earth diamonds would not be terribly strong, as there would be so many opportunities for an attacking monster to cleave through the diamond by hitting it at the wrong angle. This does not happen here on Planet Minecraft, though. Here, the strength of the diamond maintains itself at the highest level just as well on a chest plate as it does on an ax. For that same reason, you can even make a sword out of diamonds, which would also not be terribly practical with diamonds from Earth.

Oh, and one more thing. While it is true that a day's worth of mining may only yield a dozen or so diamonds, it should be noted that these diamonds are, in purely scientific terms, flipping enormous. When you mine a block of diamond ore, it always yields exactly one identically sized diamond. Placing nine of these diamonds into the crafting table gives you a block of diamond that measures exactly one cubic meter (35 cubic feet), which, assuming Minecraft diamonds have the same density as Earth diamonds, weighs approximately 3,514 kilograms (3.9 tons). This means that each individual diamond weighs approximately 390.4 kilograms (861 pounds). Now, because diamonds on Earth are so small, they are pretty much always measured in carats rather than kilograms, with a single carat being equal to 0.2 grams. You can probably see where I'm going with this. . . .

If there are five thousand carats in a single kilogram, and each Minecraft diamond weighs 390.4 kilograms (861 pounds), then every single diamond on Planet Minecraft is a 1,950,000-carat diamond.

For reference, the largest rough diamond ever found on Earth was a bit more than 3000 carats, and that has a value of about $400 million.

GEOLOGIC RANDOMNESS

The last subject I'd like to discuss in this chapter has to do with the location and placement of various rocks and minerals on Planet Minecraft. This is, without a doubt, one of the most strange and puzzling mysteries I have yet come across since my arrival here, and while I feel the need to present my findings on this matter, I regret to say that I still do not have anything approaching a coherent theory as to how it can be explained.

The best way to explain this mystery is by talking a bit about andesite and coal. Now as you probably know, both andesite and coal are relatively common back on Earth, just as they are relatively common here on Planet Minecraft. Furthermore, the andesite and coal that you find here both display many of the same properties as the andesite and coal you will find on Earth. On both planets, andesite is a fine-grained gray rock speckled with pale, yellowish crystals and coal is black rock speckled with shades of dark gray. Minecraft andesite is hard, strong, and durable just like Earth andesite, and Minecraft coal is softer than andesite and combustible, just like Earth coal is. In fact, there is not a single quality about either of these two rocks that is significantly different between the two planets, except, of course, for the ways that everything is different on the two planets.

So what, exactly, about these two rocks has me so mystified? Simple: their location.

Allow me to explain. On Earth, andesite is what is known as an extrusive igneous rock, which means that it is formed when magma exits the Earth and cools either above the surface or at least very close to it. Coal, on the other hand, is an organic sedimentary rock, which means that it forms when dead plants accumulate in an oxygen-deficient environment like a swamp and then get compressed for millions of years. The circumstances and environment necessary for the formation of these two rocks literally could not be more different—exactly as different, in fact, as the difference between a lake of boiling hot magma and the bottom of the cold, dank swamp. If there's one thing you could say for certain about these two types of rocks it is that they could never, ever form alongside each other. Not only are their required environments so incredibly disparate, but if the magma necessary for the formation of andesite were to ever get anywhere close to coal, or any of the building blocks of coal, it would literally incinerate every last ounce of it, because coal, remember, *is flipping combustible!* Seriously, finding coal next to andesite on Earth would be like finding ice cream in a convection oven.

Yet, here on Planet Minecraft, that's exactly where I find coal. Frequently, in fact. The first time I ever found andesite here, I was looking for coal, and the coal was actually surrounded on every side by andesite. That, my friend, is the great mystery of the geology of Planet Minecraft.

It's not just coal and andesite, either. From what I can tell, there seems to be practically no rhyme or reason to the location or formation of almost any of the rocks and minerals beneath the surface of this planet. I mean, sure, there are a few things that form in more or less comprehensible ways (like obsidian forming beside lava, for example) but the vast majority of them just seem to be randomly placed in blobs under the ground. It's almost as if someone took a normal planet, threw it in a giant blender, and then just pressed everything together and planted some grass and trees on the surface.

So, yeah—that's probably about the most reasonable theory I'm going to come up with to explain the geologic randomness of Planet Minecraft: the Giant Blender Theory.

CHAPTER 3
MINECRAFT BIOMES

W hile there are dozens of unique environmental areas, or biomes, here on Planet Minecraft, I am going to limit my discussion in this chapter to the hottest and coldest of them, as these are the ones that offer the most unique perspective into the environment of this strange planet. Since I began my research and made my home in the badlands, that's where I will begin this chapter as well.

BADLANDS AND DESERTS

Much like on Earth, the dry biomes consist of areas with bright sunlight, little to no precipitation, and sparse vegetation. Of course, on Earth, it does rain occasionally even in the driest of deserts, but I did not observe any rain whatsoever in any of these biomes. Presumably, any water used by the sporadic plant and animal life comes from what must be plentiful underground sources. When water does appear in these underground sources, usually bubbling up in small ponds and streams, much like Earth there is a proliferation of vegetation immediately surrounding the oasis.

BADLANDS

I'll start by discussing the badlands, as it's where I finally decided to make my home on this world. At first glance, the badlands here in this world appear quite similar to the badlands we are accustomed to on Earth, such as those found in the western parts of the United States. As with the Earth, this biome seems to be made up mostly of clay-rich soils, with steep cliffs and plateaus (known as mesas) that seemed to have been eroded away over time by the force of wind, rain, and water. While there are many colors to these sedimentary rock cliffs and ravines, they tend to be varying hues of red and yellow, making them quite beautiful indeed.

Okay, let's get down to some of the science. On Earth, badlands are formed by two relatively simple geological processes: deposition and erosion. Deposition is the process in which various sediments are deposited in layers on top of each other over millions and millions of years. Often this happens as a result of water carrying silt, sand, and other loose materials and depositing them evenly at the bottom of the body of water, but it can also happen as a result of wind, glaciers, etc. Basically, it's like making one of those art projects where you drop different colors of sand into a clear glass bottle, layering each one on top of the other to create a rainbow effect. Each time the feature on the surface changes, for example, from a large lake to a frozen glacier, different types of material will be carried into the area and dropped on top of whatever was there before. This causes the layer cake effect that we see quite easily both in badlands formations on Earth as well as here on the Minecraft world. Once these layers have been formed one on top of the other, they become ripe for erosion.

Erosion is simply the process of rocks (or other material) being slowly worn away by flowing water, wind, glaciers, etc. For example, when a fast-moving body of water flows over soft layers of deposited minerals over the course of millions of years, the soft minerals slowly wear away to reveal a cross-section of the layers beneath the surface.

Think about it this way: Imagine you make a cake with many different types of layers. Starting from the bottom, you've got chocolate cake, followed by vanilla frosting, followed by vanilla cake, followed by chocolate frosting, followed by strawberry cake, followed by strawberry frosting . . . you get the picture. Then imagine that you take your cake and put it under a faucet so that the water falls on one side of the top of the cake and flows down off of the other side. If you leave the water running for a few hours, what will happen? Just like in nature, each of the layers will wear away, one after the other until you have a canyon running down the middle of your cake. Looking at the sides of your cake canyon, you would be able to see each layer of the cake just as you can see each layer of minerals in the badlands.

That's how it works on Earth, at least. But what about here in the Minecraft world? Well, there are a few reasons to assume that the badlands here must have formed in much the same way that they formed on Earth. The main reason for this is simply the clear existence of layered soft clay-like mineral formations resembling the mesas, plateaus, and cliffs of the badlands on Earth. Furthermore, one can occasionally see examples of geologic features called hoodoos around the badlands. A hoodoo is a tall skinny tower of rock that is frequently found in badlands and similar types of geologic regions. These are formed on Earth when a hard mineral, such as granite, forms on top of a number of layers of softer material like clay. This heavy, hard material presses down on the softer material beneath it, compressing the soft material and making it stronger than the surrounding clay that does not have such a heavy piece of rock sitting on top of it. Then when erosion takes place, this column of compressed, dense clay wears away at a significantly slower pace than the soft material around it, creating the tower known as a hoodoo.

Again, the mere existence of these types of formations suggests to me that this region formed in a similar way that such regions formed on Earth. However, given the significant differences in basic physics like gravity (see my later chapter on gravity and physics for a discussion of this), one does have to leave open the possibility of other causes. For example, while I was able to find quite a few hoodoo-type formations throughout the badlands, strangely, none of them had the characteristic layer of dense, heavy rock at the top. Instead, they were simply towers of soft clay much like the rest of the mesas and plateaus in the Badlands. This isn't even the strangest thing, though, as I also saw at least two examples of what appeared to be hoodoos that were simply floating above the ground without touching anything on the surface whatsoever. While this obviously has something to do with the strange way that gravity works here (as mentioned in the chapter on gravity), the fact that these exist in this fashion at all makes it quite difficult to say for certain how they were formed.

DESERTS

I have come across a few areas of what I would call pure desert here in this world, which seem to be mostly like similar regions on Earth. They are dry, covered almost completely with sand and sandstone, and contain sparse vegetation that is limited to cacti and the occasional dead shrub. While there are some areas of desert where you can find scattered lakes or low rolling hills, most of the desert that I have come across is relatively flat, without the rocky, cliff-like geologic formations found in the badlands.

In Earth science, we generally classify a desert simply by the amount of rainfall that it receives: If a region receives less than ten inches of rainfall in a year—*bam*, it's a desert. Obviously, I have not been here nearly long enough to make that kind of a calculation but based upon the amount of vegetation in the desert here it seems quite clear that this biome meets that criteria. On that note, however, the badlands most likely also would meet the criteria of a desert, at least in terms of precipitation. The pure desert as I'm referring to it is closer to what we think of as a sand dune desert on Earth.

Back on Earth, a sand dune desert is formed through quite an interesting and unique process. Let's say you have a nice grassy plain somewhere. The first thing you do is you take away all the water. No rain, no rivers, no lakes, no irrigation—nothing. The first thing that happens, obviously, is that all the plants will die. After that happens, the soil, now devoid of roots and other plant life to hold it together, becomes much more susceptible to erosion by the elements. Over time, this top layer of soil is eroded away. At this point, the air starts getting drier and drier along with the ground. When the air gets dry enough, you start to find huge fluctuations in temperature occurring between day and night. I know, I know: you usually think that the desert is hot all the time. But it's actually not. The desert is only very hot during the daytime when the sun is beating down on the land. Once the sun goes down, there is no moisture in the air to retain the heat from the sun and all of that hot air rises instantly up into the upper atmosphere, leaving the air closer to the ground quite cold. This fluctuation between hot and cold causes the rocks of this region (the top layer of which, by this point, has been exposed to the air by the erosion of the soil) to start to crack and fall into smaller and smaller pieces. Now, these smaller and smaller pieces are similarly exposed to fluctuations in temperature, while also being small enough to get blown around by the wind and knock into each other. As this happens, the fragments of rock get smaller and smaller until you're left with nothing but sand. Once you have nothing left but sand (which is really just pulverized rock), the sand gets constantly blown around by the wind, forming the type of rolling sand dunes we associate with this type of desert.

Once again, at first glance, the deserts here look much like the deserts in southern Arizona or on the Arabian Peninsula and would seem to have been formed by a similar process. However, there are a few key differences that call this assumption into question. First of all, there doesn't really seem to be much in the way of wind here. As far as I can tell, the only way to get the sand to move at all is to either physically pick it up and carry it, or to remove the sandstone beneath it, in which case it will fall down to whatever lies below. How does a desert like this form without wind? I have no idea. But it certainly wouldn't work that way on Earth.

The other interesting fact aspect of the deserts on this world has to do with the rain. While I have never observed any rain whatsoever falling on the actual sand of the desert, I have often seen rain falling in extremely narrow bands directly over the rivers and streams in the desert. This is quite strange, to say the least. The only reason I can possibly think that this would even happen would be if the rain here came only from the evaporation of water directly beneath the rain cloud and, perhaps without wind, doesn't move around very much. Even this seems like a long shot, though. In any case, it is a definite departure from the way a desert like this would work on Earth.

SAVANNA

Much like Earth, many of the biomes immediately adjacent to the desert and badlands are a dry grassy region we would call savanna. These dry regions are characterized by limited vegetation consisting mostly of grasses and sparse trees and form a transitional zone between the desert and the forest. Unlike the savannas on Earth . . . well, actually, pretty much everything else is unlike the savannas on Earth.

First of all, savannas on Earth are formed primarily as a result of having rainfall that is greater than that in the desert, but not enough to form the more richly vegetated areas like forests or jungles. Often, this takes the form of highly variable rainfall, where it may rain a great deal but only for a very short period of each year, or the rainfall may not even come every year. In any case, there does need to be some rain to make a savanna: that's where all the grass and trees come from.

However, in Minecraft world, I didn't see any difference at all between the amount of rain that fell in the savanna, and the amount of rain that fell in the desert. Which leads to the question: How the heck are all these grasses and trees growing? Yes, the savanna does also have that strange phenomenon of rain falling directly over a body of water, but of course, this by itself is not enough to differentiate it from the desert.

SNOWY BIOMES

Far to the southeast, I came across a number of fascinating and unique biomes that seem to be in a state of perpetual winter. Perhaps the most interesting part about this region was where it is located. I had been cataloging the scarce flora and fauna of a particularly expansive section of the desert when I came across a wide body of water too large to see across. Having never ventured that far before, I decided to have a look at what was on the other side, assuming that it would probably just be more desert or savanna. So, using my homemade pocket boat (I will never get over how cool it is that a boat can fit in my pocket), I set out across the water. After a mere few hundred meters, I was quite surprised to see great towers of ice and snow rising above the horizon. As I got closer, I realized that these were in fact enormous icebergs, floating just off the coast from what seemed to be an expansive tundra.

Now, I should hope that I don't need to tell you that such close proximity between an arid desert and an iceberg-laden tundra simply cannot exist on Earth. You see, there is a fairly short list of conditions that can create such huge variations of climate like that on Earth. The first one that everyone thinks of is latitude—or, the distance from the Earth's equator—with colder climates occurring the farther away from the equator you get. Why does this happen? Well, think about it this way: Imagine you are pointing a flashlight at a soccer ball. Like the Earth, the soccer ball is spherical, which means that every point on that soccer ball will receive different levels of light. The part closest to the flashlight will receive the most direct light, falling straight down on top of it, while the opposite side will receive no light at all.

Now, if the Earth and the sun both stayed perfectly still all the time, you would have one particular point on Earth that received the most light all of the time, and hence was the hottest all the time. On the exact opposite side of the Earth, you would have the darkest point on the globe, which would also be the coldest. However, neither the Earth nor the sun ever actually sit still; not even for a second. The Earth is constantly rotating on its axis while simultaneously orbiting around the sun (which, in turn, is constantly orbiting around the center of the galaxy).

Let's pretend for a second, though, that the Earth did not orbit around the sun, but merely rotated on its axis while its placement in relation to the sun remained constant. In this scenario, there would be a ring around the circumference of the center of the globe, constantly going through a day-night cycle and always getting the most light out of any other part of the sphere. This ring is essentially what we call the equator. The farther you get toward the top or the bottom of the sphere (i.e., the North or South Pole, or the actual axis on which the sphere rotates), the less direct the light reaches the surface, and therefore the less heat is absorbed, resulting in a colder climate. This is why the North and South Poles are so cold on Earth, and why the tropical and desert regions around the equator are so hot.

It is important to note that the orbit of the Earth around the sun also plays a big part in this equation. You see, the North and South poles of the Earth are not directly perpendicular to the Sun, but rather they tilt at about 23.5 degrees away from parallel. For half of the Earth's orbit, the North Pole is tilted more toward the sun, while the South Pole is tilted toward the sun for the other half of the year. When the half of the Earth you are standing on is tilted closer to the sun, it absorbs more heat, and we call this summer. When the half of the Earth that you are standing on is pointed away from the sun, it absorbs less heat and we call this winter. This is also why the seasons are always opposite in the Northern and Southern Hemisphere (and why the days are always longer in the summer and shorter in the winter).

This brings us to the first major reason why this particular biome, or at least the placement of this biome, is so improbable and would be impossible on Earth. In order to travel from the hot, arid desert to the snowy, frozen tundra, I only had to travel west (which is to say, in the same direction that the sun moves across the sky here every day) for a very short distance. Traveling due west means that I was not in any way changing the distance from the equator (assuming this world has an equator, which is a question for another chapter . . .) and thus the angle of the sun cannot be the cause for this drastic change in climate.

So, what else causes differences in climate? Well, the next most obvious reason would be elevation. I'm sure you have noticed on Earth before that the top of mountains always seems to remain covered with snow far longer than the valleys beneath them. The reason for this is fairly simple. You see, the air that surrounds us all on Earth is in fact made up of trillions and trillions of tiny little particles of gas. On Earth, this gas is made up mostly of nitrogen with some oxygen, carbon, and other things thrown in as well. Now, like everything else in our universe, each and every one of these individual molecules of gas are affected by gravity and fall toward Earth until they run into something heavier or denser (for example, the ground, or more densely packed gas particles) and can't go any farther. As a result of this, the heavier gas molecules collect closer to the surface than the lighter gas molecules, and the gas molecules closer to the surface also have to support the weight of all the molecules above them, compressing them even farther. All of this leads to the air near the surface of the Earth being much denser than the air higher up in the atmosphere.

Now, while the sun is obviously the source of pretty much all of the heat on the surface of the Earth (not counting thermal heat, of course), the sun actually has to heat up *something*: water, land, air, etc. So, when you are at a very low altitude, for example, sea level or below, the air is very dense and so the sun has lots of particles to heat up. Not only that, all of these particles of gas hold on to the heat and help to spread it around to everything on the surface and immediately above the surface. When you travel to higher and higher altitudes, however, the density of the air decreases (meaning, literally, that the number of particles of gas per square inch is lower) so there are fewer things to actually absorb that heat from the sun and transfer it to anything else. When combined with the effect of latitude, you can easily see how climatic extremes occur on Earth. When you have a low-lying area that is close to the equator, it will always be very hot and when you have a high-altitude mountainous region far from the equator (like the Rocky Mountains in Alaska, for example) it will always be relatively cold.

Once again, this common reason for climatic variation on Earth does not apply here in the Minecraft world. The glaciers and icebergs I found here were floating directly at sea level, just a short trip across the water from the desert. The only place where anything even remotely like that happens on Earth is in the Antarctic Ocean. There are a few places in southern and western Australia where you could theoretically get on a boat in the desert (or semi-arid grasslands that are very close to being a desert), sail in a straight line away from the coast, cross the Southern Ocean, and the first solid ground you would find would be the frozen glaciers and icebergs of Antarctica. However, the closest two points on Earth where you can do this would require a trip of nearly 4000 kilometers (2485 miles) (and more than 30 degrees of latitude) which is obviously much farther than I traveled between the desert and the tundra here.

The other major cause of significant climate variations on Earth is the ocean. Since the Earth is made up primarily of water, it is the ocean that absorbs the vast majority of heat energy from the sun. This heat energy is then transferred into the air by the evaporation of water into a gas, and as this gaseous water vapor mixes with the air in the region it may warm or cool it depending on whether the ocean temperature is hotter or colder than the temperature of the air. Now, in a stagnant or fully enclosed body of water, this would not greatly affect the climate on land, as the water and the air over land would be absorbing relatively consistent amounts of heat throughout the year. However, the oceans of Earth are not stagnant at all. In fact, ocean currents keep the liquid cover of Earth in a constant state of motion, bringing warm water from the equator toward the poles and then cool water back from the poles to the Equator.

The air over the landmasses that sit beside the ocean currents that move away from the equator and toward the poles, therefore, winds up getting warmed by the huge amounts of warm water passing by the coastline.

Similarly, landmasses that sit alongside the ocean currents that bring colder water from the poles down toward the equator will be cooled by the ocean. This is also why you will often find that places on the immediate coast of the ocean will have less extreme temperature variations over the course of a year than places farther away from the ocean. In the summer, when the sun is heating the air over land up to a temperature higher than the ocean, the air will become cooler the closer you are to the ocean itself. In the winter, when the air over land has dropped in temperature below that of the ocean, the air will be warmer the closer you are to the ocean. Thus, if you were to travel 100 kilometers (62 miles) directly away from the ocean during the summer months in New England, for example, you would almost always find it to be a few degrees hotter, while if you were to take the exact same trip in the winter, you would almost always find it a few degrees colder.

Once again, however, the existence of a frozen climate in such proximity to that of a hot desert climate in this Minecraft world cannot be explained by ocean currents. For one thing, the body of water between the two landmasses is just not nearly big enough to account for such a difference. Even if it were an enormous body of water, though, there is no scenario wherein water temperature and currents alone could account for this kind of variation in climate in the absence of factors like latitude and elevation.

So then, what does allow these vastly different climates to exist in such close proximity to each other? The short answer is that I have no idea—but I do have a theory. You see, there are two things that I do know about this planet: there is a whole lot of volcanic activity, and very little wind. Now, volcanoes on Earth release a whole lot of different gases into the atmosphere, some of which (like sulfur dioxide) can actually create a climatic cooling effect. This happens when sulfur dioxide from a volcano gets shot way up into the stratosphere and rapidly cools into sulfuric acid vapor, which, in high enough concentrations, can actually reflect some of the sun's heat back into space. At the same time, volcanoes can also release large amounts of carbon dioxide into the atmosphere, which, in high enough concentrations, actually traps the sun's heat close to the surface. This is known as the greenhouse effect and causes the temperature on the surface to rise.

As I mentioned, there is basically no discernable wind on this planet, which means that, in theory, any gases that are hovering over the surface of a particular region very well might stay right there in that region. If (and it's a big if . . .) there was some kind of volcanic activity that occurred a long time ago which caused high concentrations of sulfur dioxide to sit above some regions and high concentrations of carbon dioxide to sit above other regions, it is plausible that the result would be a significant climate variation that was not attributable to latitude, elevation, or ocean currents. Of course, nothing like that has ever been observed on our planet, but you could say the same thing about cubic chickens and floating trees.

Aside from the location of these frozen biomes, they do mostly coincide with similar biomes on Earth. They are characterized by snow-covered hills and mountains, sparse vegetation consisting mostly of spruce trees and the occasional small shrubs, with the only animal life that I could see being little white rabbits and large white bears. I won't call them polar bears because I can't really say that we are anywhere near a pole of any kind, but they certainly do look like polar bears.

One more interesting characteristic worth noting about this region are the formations that look like giant towers of pure ice rising out of the ground. I was hiking near the center of this frozen landscape when I first came across these magnificent formations. Covering a huge section of the terrain, these ice towers rose almost every few meters from the snow. There are similar formations on Earth, but with some important differences. Known as fumarolic ice towers, these structures only form on Earth above areas of significant volcanic activity. Any place with significant volcanic activity can have a fumarole, which is simply when a fissure forms above a pocket of volcanic gases, allowing the gases to escape as if from a chimney through the surface of the Earth. When this happens in an extremely cold climate, like Antarctica for example, the hot volcanic gases will melt the frozen water and snow at the top of the fumarole, instantly boiling it into water vapor which then rises before quickly freezing in the frigid air above. Over time, this constantly rising and freezing water vapor forms a tower of ice that can look somewhat similar to the ice towers I found here in the Minecraft world.

As I mentioned, though, there are a few very important differences between the ice towers here and the ones back on Earth. First of all, fumarolic ice towers always have gas pouring out of the top of them, as they are essentially ice chimneys, and they rarely form in numbers anywhere near the thousands I found here. More impressively, though, is the issue of the height of these ice towers. On Earth, ice towers are usually around 6 to 7 meters (20 to 23 feet) tall and almost never reach heights over 10 meters (33 feet). In the Minecraft world, however, some of these towers reach heights surpassing 40 meters (131 feet) or more, with some of the larger ones perched atop tall mountains even climbing high enough for their tips to touch some low flying clouds.

CHAPTER 4

MINECRAFT PHYSICS

My mission here is primarily to explore, discover, experiment on, and note the differences between this strange world of Minecraft and our home planet of Earth. Now that I have secured shelter, food, and the other basics of survival, I want to get right down to the experiments, and there couldn't be a better place to start with my significant research than gravity.

GRAVITY IN MINECRAFT

As a scientist, one of the first things I noticed as being a little bit off about the Minecraft world is its gravity. A *little* off is probably the understatement of the millennium—gravity in Minecraft is downright looney tunes. Before we get into all the different ways that gravity is different here, I should start with a basic explanation of what gravity is in the first place.

You probably heard the story as a child: The great scientist Sir Isaac Newton was just sitting around, minding his own business, when he felt something strike him right on the top of his brilliant head. He looked down to see what the offending object was and saw an apple lying on the grass beside him. Looking up, he noticed that he was sitting under an apple tree and deduced (remember: the man was a genius) that the apple must have fallen off the tree. All of a sudden, somewhere deep in the caverns of his nimble brain, a light bulb went off (or would have, if light bulbs had been invented at that time in history) and a thought popped into the forefront of his consciousness: *Some invisible force must have made that apple fall on my head.* So off he marched to sequester himself in his study, and before you know it: Newton discovered gravity.

Of course, that's almost certainly not how any of those events actually transpired, but it's a nice story. If I remember the actual history correctly, Newton was in fact inspired by seeing an apple fall from a tree, though that event only took place long after he first began his research into gravity. Furthermore, this event did not inspire Newton to question why the apple fell from the tree, but rather why two apples on opposite sides of the Earth would fall in opposite directions (which is to say, they would both fall toward the center of the Earth), and whether the answer to that question could also answer his questions about why the moon revolves around the Earth, and why the Earth revolves around the sun.

As it turned out, Newton's suspicions on that fateful day in 1666 were quite correct and answering them led him to the formation of his world-changing Universal Theory of Gravitation.

You see, by that point in history, humans had long known that the Earth was a sphere and that some invisible force kept everything on its surface from floating up into the heavens. What Sir Isaac Newton worked out (quite ingeniously) was that the movement of every celestial body in the sky could be explained and predicted if you just assumed that all of them created the same force that made the apple fall from the tree and that this force was actually strong enough to affect the movement of objects far, far away from the celestial bodies themselves.

To test this novel idea, Newton came up with a brilliantly simple thought experiment: Let's say you place a cannon at the top of a tall tower and fire it in a direction exactly parallel to the ground. When the cannonball first exits the barrel of the cannon, it will be traveling parallel to the ground, of course, but it will very shortly start to fall to the Earth as its momentum is slowed and eventually stopped by that mysterious force that keeps everything from flying off into space. The actual distance the cannonball will travel depends upon many factors, of course, but Newton knew that identical cannonballs fired from the same cannon with the same amount of powder at the same angle will more or less travel the same distance every time. Furthermore, he knew that the higher you made your tower, the farther the cannonball would travel before the invisible force brought it to a resting position on the surface of the Earth. The question that Sir Isaac Newton asked in this thought experiment was this: If there were no limit to the height that you could build your tower, would it be possible to build one tall enough that a cannonball fired from the top of it would travel all the way around the Earth and not touch the ground until it wound up back where it started? Furthermore, if that were possible, would it also be possible to fire a cannonball from a tower so high that it would never reach the ground at all, but rather just keep on circling the globe indefinitely or else just fly off into space?

Of course, the answer to all of these questions (at least theoretically) is yes.

Now, once Newton had worked out that such an experiment would theoretically succeed, he applied that same idea to the moon, the sun, and everything else up in the heavens, and quickly realized that it explained the motion of the heavens far more effectively than any theory that anyone had ever thought of before.

This theory, which he aptly named his Universal Theory of Gravitation, remained humanity's best way of understanding the mechanics of the universe for nearly five centuries. In fact, it wasn't until a young physicist named Albert Einstein came along with his groundbreaking Equivalence Principle that humanity finally learned that there was far more going on with gravity than what Newton deduced back in the seventeenth century.

So, what was Einstein's Equivalence Principle all about? Well, it does get rather complicated, but the simplest way of thinking about it is this: there is no functional difference between gravity and acceleration.

I know, I know . . . I said it would be a *simple* way of thinking about it. Let's try another thought experiment.

Imagine you are standing in a room on the surface of the Earth, holding an apple in your hand with your arm stretched out straight in front of you. You drop the apple, and time how long it takes to reach the floor, finding that it takes exactly 0.55 seconds. If you were to conduct the same experiment, but this time in a stationary space capsule floating far from any source of gravity, the apple, obviously, would never touch the floor at all.

With me so far? Good.

Now, let's do the second experiment one more time, but this time we do it while the space capsule is accelerating at exactly 9.80665 meters per second squared in a direction exactly perpendicular to the floor of the capsule. How long will it take the apple to reach the floor this time? If you guessed that it would take exactly 0.55 seconds, then you, my friend, would be correct. This is what Einstein meant when he said that there is no functional difference between acceleration and gravity. It just so happens that the gravity we experience on Earth is equal to an acceleration of 9.80665 meters (32.1737 feet) per second squared. If you were to try the same experiment on a different planet, you would wind up needing a different rate of acceleration to make the apple reach the floor of the space capsule in the same amount of time it takes for the apple to reach the floor on the surface of the planet.

Now we're going to make things a little more complicated and a lot more interesting. You see, Einstein's discovery of the Equivalence Principle was really just a stepping-stone to a much bigger discovery.

Not being satisfied to simply understand that acceleration and gravity were functionally identical, Einstein wanted to figure out *why* they were functionally identical. Answering this question was what led Einstein to develop his world-changing Theory of Relativity . . . but we'll save that story for another chapter. For now, it's enough that you understand the basics of Newtonian gravity, and how Einstein expanded on it by discovering that acceleration is equivalent to gravity.

FALLING

Okay, okay. Now that we've got the science lesson done, let's get on with the experiment. The first experiment I wanted to conduct was simply to see how fast objects fall in Minecraft and compare that to how fast they fall on Earth, just to get a baseline understanding of the difference between this planet and our own. As Einstein taught us, the force of gravity on the surface of Earth is approximately 9.8 meters (32 feet) per second squared. That said, even on Earth, gravity isn't exactly consistent everywhere on the planet. Because gravity is a product of mass, the parts of Earth where there is greater mass beneath the surface (for example, in places where there are large deposits of heavy metals beneath the ground) will have slightly stronger gravity than the rest of the planet, though the difference will never be strong enough to feel. The fact that the Earth is constantly spinning also affects gravity, as this rotation causes a certain amount of centrifugal force acting upon everything on the surface of the Earth. Yes, this means that the spinning of the Earth is actually trying to push you off the surface of the Earth, while gravity is pulling you down toward it. Obviously, gravity always wins. That said, because the spinning is faster at the equator than at the poles, the centrifugal force is higher, and the gravity is actually lowest at the equator. But because gravity is only 0.043 meters (0.14 feet) per second squared less at the equator, it still doesn't make enough of a difference to notice.

So, right off the bat, I'm just going to go with the assumption that whatever the gravity is on the Minecraft world, it is fairly consistent everywhere. Given the crazy nature of gravity here, this might not be an accurate assumption, but it's what we're going to go with to start.

One more note: due to this planet's clear lack of hi-tech labs, recording equipment, etc., I have to rely on simply dropping small things off of tall things and timing how long they take to reach the ground in order to calculate gravity. This does not take into account the impact of air resistance, which is obviously quite important to determining actual scientific data, but as far as simply comparing the gravity of Planet Minecraft to Earth, it's easier just to conduct these experiments as if they were in a vacuum.

Anyway, as mentioned before, the force of gravity on the surface of Earth is about 9.8 meters (32 feet) per second squared. This means that an object dropped toward the Earth will fall at 9.8 meters (32 feet) per second after the first second, but 19.6 meters (64 feet) per second after the second second, 29.4 meters (96 feet) per second after the third second, etc.

Using these calculations, I worked out that an object dropped from a height of 100 meters (328 feet) on Earth would take approximately 4.5 seconds to reach the ground, at which time it would be traveling at approximately 44.25 meters (145 feet) per second. An object dropped from a height of 200 meters (656 feet) would take approximately 6.4 seconds to reach the ground, at which time it would be moving approximately 62.6 meters (205 feet) per second.

Now it's time to compare! Ideally, I would have some kind of a setup for this experiment involving a large tower, a simple object, a trigger mechanism to drop that object, and a stopwatch that automatically starts when the object is dropped and stops automatically when it hits the ground. I didn't really have any of those options at my disposal at the time, however. So, I just decided to build some big old towers and jump off them. Don't worry, though: from what I had already learned, you can fall from pretty much any height on this planet and be just fine, as long as you fall into at least one block of water. This is, of course, completely insane (and should probably be the subject of some future experiments), but for the time being, I decided to use it to my advantage, and built some big towers right in front of a small pool of water. Also, because I could not for some reason build a tower taller than 256 meters (840 feet) in the air, I decided to dig a big hole in the ground and put my pool in the hole so that I could get a bit more height for the experiment.

Well, digging a hole deep enough wound up taking longer than the rest of the experiments combined, but after a few days of digging I was finally ready to start my experiments. I had one tower at 200 meters (656 feet) tall, one tower at 100 meters (328 feet) tall, one tower at 50 meters (164 feet), and one tower at 25 meters (82 feet).

I finally managed to time myself falling off each of the towers, and here are my results. Note that, because there is some human error involved in the stopwatch side of things, I actually jumped off each tower ten times and averaged the results. Here is what those same fall times would be in Earth gravity of 9.8 meters (32 feet) per second, and what they averaged in Minecraft.

Tower Height	Earth Fall Times	Minecraft Fall Times
200 meters (656 feet)	6.4	4.4
100 meters (328 feet)	4.5	3.1
50 meters (164 feet)	3.2	2.1
25 meters (82 feet)	2.3	1.5

Now, using the same math that we use to determine gravitational acceleration on Earth, I determined that the gravitational acceleration on Planet Minecraft is approximately 21 meters (69 feet) per second squared. That's more than double the gravity on Earth! Just for some context, Jupiter, as the largest planet in our solar system at more than eleven times the size of Earth and 317 times its mass, has a gravity of 24.79 meters (81 feet) per second.

What does all this mean? Who cares if the gravity is greater here? What's the point? Well, there are a number of implications to this. First of all, gravity this high, in theory, should cause a lot of things to happen here that frankly don't appear to be happening. For one thing, everything in this world should weigh a little more than twice as much as it does on Earth. So, for example, iron weighs approximately 7,873 kilograms (8.7 tons) per cubic meter on Earth. That would mean on Planet Minecraft that same cubic meter of iron should weigh nearly 17,000 kilograms (18.7 tons). That's roughly the same weight as eight Ford F150 pickup trucks stacked on top of each other. Now, unfortunately, I don't have a scale here to measure weight, but just for fun I decided to see how much iron I could carry and still manage to walk. Well, with thirty-seven pockets each containing 64 cubic meters (2260 cubic feet) of iron, on a planet with a gravity of 2.1g, I managed to easily frolic around the landscape here while carrying a bit more than 38 million kilograms of metal, and I never even broke a sweat. Heck, I even went for a swim!

This brings us to the real conundrum of gravity in this world, and it's a doozy: Gravity simply does not affect everything equally.

DENSITY

That gravity does not affect all things in the same way should hardly come as a surprise to anyone who has visited this world, as one of the first things they likely did was punch out a nice meter-high section of a tree trunk, only to find that the tree just kept on floating in place instead of falling to the ground.

Actually, floating might not even be the best word to describe what happens to the trees in this scenario, as they don't behave at all the way that floating objects do on Earth. On Earth, objects float when the density of the object is equal to or less than the density of the fluid the object is suspended in. Another way of thinking of this is that the upward force of the fluid (known also as its buoyancy) is equal to or greater than the downward force of gravity. Now, before you get all worked up and try to tell me that balloons float and they are in the air, remember that fluid is not the same thing as liquid. Gases are fluid just as much as liquids are fluids. Picture a flat wooden board floating on a still body of water. Why doesn't the wood sink? Well, as we mentioned, because the density of the piece of wood is less than the density of the water, meaning that the force of gravity acting on the piece of wood is less than the force pushing up on the wood by the water. With me so far? Good. Now, what would happen if you were to add small lead weights on top of the wood one at a time? As you can probably guess without having to try the experiment yourself, the wood will continue to float until the combined density of the weights and the wood surpasses the buoyancy of the water, at which point it will sink (this, of course, is because the lead weights are far denser than either the wood or the water).

Using this logic, if a tree is floating on Planet Minecraft after removing its bottom block, you should be able to get it to stop floating by adding enough weight on top of it. This is a fairly simple experiment, so I decided to give it a try. I walked out from my dwelling, found myself a nice small oak tree (making sure to find one all by itself, so it wouldn't be affected by touching any other trees), and punched out its bottom block. As always, the block popped right off into a miniature version of itself, while the rest of the tree remained motionless. I then built a small staircase adjacent to the tree, climbed on top, and began stacking blocks of granite directly over the trunk. The first block of granite, predictably, did nothing. So I tried some more. Ten blocks? Still nothing. Not wanting to leave any room for doubt, I stacked 100 blocks of granite on top of the tree, and . . . still nothing.

Now on Earth, a square meter of granite weighs approximately 2,691 kilograms (1.3 tons). This means that adding 269,100 kilograms (135 tons) to the top of the tree still did not cause the force of gravity acting on the tree to surpass the upward force of the air the tree was floating on.

Now I know what you're thinking: I just said that buoyancy was a product of density, not weight. And that is very true, obviously. If weight were the only factor in determining buoyancy, then it would make no sense for a huge cargo ship to float on the ocean while a bowling ball sinks right to the bottom. In reality, however, there is enough air inside the cargo ship to decrease its total density enough so that it is less than that of water. Thus, the cargo ship floats, whereas the bowling ball, being solid, does not.

Now, doesn't all this mean that the granite itself might just be less dense than the air on Planet Minecraft? Of course it does (as ridiculous as that sounds). I wasn't actually sure about this at the time, though, so I went ahead and stacked two granite blocks on top of each other and then removed the bottom one. What happened to the granite on top? It stayed right where it was, of course. So, does this mean that the air on Planet Minecraft is far more dense than the air on Earth? Or rather, is it that the granite on Planet Minecraft is far less dense than the granite on Earth? Well, it could mean either, or both (or more likely, neither, but we'll get to that possibility a little later). But for the purpose of my experiments, if I wanted to find out what it would take to get the tree to stop floating, I would need to stack something on top of it that does not float.

The first thing I tried, obviously, was myself. I do not float, after all, so it stands to reason that I am denser than the air on this planet, and therefore I could increase the overall density of the tree just by standing on top of it. Alas, when I tried standing on top of the tree, nothing happened at all—it still just stayed exactly where it was. I even tried jumping up and down a bit to see if I could get the tree to move, but it wouldn't so much as budge.

Okay, I thought, *maybe I am simply not dense enough* (though I had a few teachers in elementary school who might say otherwise. . . .)

My next step was to find something else, ideally something stackable, that didn't float. This turned out to be far more difficult than you would think, as there are very few materials here that do not float when the bottom layer is removed. One of these materials, however, is sand. If you stack a few blocks of sand on top of each other and remove the bottom one, all of the others fall right down. Right away, we can see a very clear difference between Planet Minecraft and Earth: Sand here seems to be denser than granite. This was a fascinating observation, but one that I decided to explore at a later date, as my present priority was to continue my experiment and try stacking sand on top of the tree. So that's exactly what I did. One block: nothing. 10 blocks: nothing. 100 blocks: nothing. I could have gone on, but there really didn't seem to be much of a point. The sand was obviously not affecting the tree's ability to float.

So, what did I learn from all this? Well, it is of course possible that, while the sand is dense enough to sink in Minecraft air, it is just barely dense enough to do so, and therefore an unreasonable quantity would be required in order to increase the average density of the tree enough to make it sink down to the surface of the planet. This would have been a reasonable assumption, but when I tried the same experiment with other materials that sink on their own (such as gravel, concrete powder, and even some anvils that I made) the results were the same, and it seemed quite unlikely that all of these things are so minutely more dense than air that I could create enough of them to affect the flotation of the tree.

This is when something else occurred to me. Going back to our example of stacking lead weights on top of a floating wooden board, it occurred to me that, even before enough weight has been put on the board to sink it, the board will still move up and down every time a weight is placed on top of it, as it finds its equilibrium between the two forces of buoyancy and gravity. This does not happen when stacking things on top of the tree. This was the final piece of evidence I needed to confirm what I suspected all along: something other than buoyancy must be responsible for the tree's ability to float seemingly in midair.

So, at this point, I was left with quite an interesting riddle: What can float without relying on buoyancy?

Well, the answer to that riddle is actually rather simple: nothing. The whole idea of floating requires the whole idea of buoyancy. However, floating is not the only way that something can hang in the air. Obviously, you have lots of mechanical ways of getting things to remain airborne (jets, helicopter rotors, and other things of that nature, mostly), but obviously, none of those were being used to keep my tree off the ground. This basically left me with one more explanation (or at least one more explanation that can in any way relate to what we understand from physics on Earth) and that is magnetism.

MAGNETISM

So, what exactly is magnetism?

To answer that question, we need to start by talking about some very small things, and their even smaller component parts: we need to talk about atoms and their protons, neutrons, and electrons.

If you have ever seen an illustration of an atom, you will probably imagine it looking like a tiny cluster of bubbles all stuck together (the protons and neutrons, which together form the nucleus) with a bunch of other smaller bubbles all circling around it (the electrons). Technically, this is not a particularly accurate representation of an atom, but it will work well enough for the purpose of understanding magnetism, so we'll just stick with it.

For the moment, let's not worry about the protons and neutrons at the center of our atom, and just focus on all those little electrons circling around the nucleus. Now, what you need to understand about electrons is that they are not just circling around the nucleus; each one is also spinning on its own, in much the same way that the Earth is spinning on its own while simultaneously orbiting the Sun. Furthermore, it's important to understand that these electrons are not just spinning any which way they choose. In fact, they can only spin in one of two possible directions: north or south.

Now each and every electron on each and every atom in the entire universe is constantly generating a tiny magnetic field, with the direction of that magnetic field pointing in the same direction that the electron is spinning—north-spinning electrons generate a north-facing magnetic field, and south-spinning electrons generate a south-facing magnetic field.

Okay, so the next important piece of this puzzle is the fact that most elements in our universe are made up of atoms whose electrons only come in pairs, and each of these pairs of electrons is always made up of one south-spinning electron and one north-spinning electron. Thus, the tiny magnetic field that each electron is generating is canceled out by its partner, resulting in the atom as a whole not generating any magnetic field whatsoever.

Of course, it's vital to note that I said *most* of the elements in our universe; not *all* of the elements in our universe. This is because a select few elements have some lonely little electrons that don't have a partner at all. That said, the majority of these elements just happen to have an even number of unpaired electrons, with half of their unpaired electrons spinning north and the other half spinning south. As the total number of north-spinning electrons on these atoms is equal to the total number of south-spinning electrons, they all still wind up canceling out each other's magnetic field, resulting in the atom as a whole still not generating a net magnetic field.

With me so far? Well, even if you're not, we're about to get to the good part, so pay attention.

There are, in our universe, exactly three elements with an odd number of unpaired electrons: iron, cobalt, and nickel (these are known as the ferromagnetic elements). Now, as you have probably already figured out, when an atom has an odd number of unpaired electrons, there will always be one electron generating a magnetic field without any other electron around to cancel it out. When this happens, the atom as a whole will be generating a magnetic field equal to its one lonely, unpaired electron. If this electron is spinning north, the atom will be generating a north-facing magnetic field, and vice versa.

Okay, now let's broaden the scope a little bit, and look at what happens when you get a whole bunch of these magnetically charged atoms floating around next to each other. The first thing you would notice if you could actually see these atoms in their natural state would be that they tend to cluster together based on their specific magnetic charge. These clusters are called domains, and being as they are made up entirely of atoms with the same magnetic charge, each domain will itself have a magnetic charge equal to its component atoms, albeit much stronger than an individual atom's charge. That said, even a tiny sliver of one of our three ferromagnetic elements will be made up of millions of individual domains, which, on the whole, will mostly cancel each other out in the same way that individual paired electrons cancel out each other's magnetic charge.

Thus, any quantity of the element large enough to see with the naked eye will usually wind up being magnetically neutral.

So then, how exactly do we wind up with chunks of these elements that are actually, you know, magnetic?

Simple: You need to get all of the domains in your chunk of ferromagnetic metal to point in the same direction. Luckily, this is a fairly simple task to accomplish.

The easiest way to get all of the domains in a given piece of ferromagnetic metal to point in the same direction is just to rub it with another piece of ferromagnetic metal that is already magnetized. The magnetic field generated by the magnetized chunk of metal will extend into the non-magnetized chunk of metal and literally pull all of the domains so that they're pointing in the same direction, thus making both chunks of metal magnetically charged. You can also accomplish the same effect by passing an electric current through your non-magnetized chunk of metal. As the electricity passes through the domains, it will pull each of them in the direction that the electricity is traveling, once again resulting in a chunk of ferromagnetic metal that generates a magnetic field.

Now we get to the last important piece of this picture: the magnetic poles. For this final piece, I want you to picture a magnetized iron rod with the word "north" written on the top and the word "south" written on the bottom. If you could see the magnetic field that this piece of iron is generating (which, of course, you can't) you would see it streaming out of the north side of the rod, extending out a little way before changing direction and sweeping down toward the south end of the rod before finally changing direction again to the north and flowing back inside the rod.

This leads us to why magnets are attracted to each other. If you take two rods like this and place one on top of the other so that the south end of the top rod is touching the north end of the bottom rod, the magnetic field flowing out from the north pole of the bottom rod won't sweep back down onto itself, but will rather get caught up in the magnetic field of the top rod and instead flow right into its neighbor's south pole, causing the two magnets to stick together. If, on the other hand, you stack them in the opposite direction, with the north ends of each rod pointing at each other, the magnetic fields will be flowing in opposition, and the magnets will repel away from each other.

This, finally, is how we get back to our problem of the floating trees on Planet Minecraft. By positioning one magnet on the ground with its north pole facing up and then positioning another one exactly on top of it with its north pole facing down, the top magnet will hang in the air, as the force of the magnets repelling each other would be greater than the force of gravity trying to pull the top magnet down.

Now before you go getting yourself too excited, this is obviously not what is happening with my tree. Even if the tree were magnetic and its pole was opposite the magnetism of the ground on Planet Minecraft, it would still be affected by weight being stacked on top of it, and it would bob around whenever somebody jumped on the top of the tree. In fact, even the slightest force applied to simple magnetic levitation like this would be enough to send the magnets repelling away from each other.

There is a specific phenomenon in magnetism known as quantum locking, which just might get us a bit closer to an explanation. In quantum locking, a superconductor is placed on top of a magnetic field, which produces the Meissner effect and causes the superconducting material to become quantum locked in place. This is essentially the same as magnetic levitation; however, it also has forces that keep it from moving around. On Earth, this is extremely hard to accomplish, and requires supercooling the magnetic materials in order to activate their superconductive properties and can really only be done in highly specialized experimental circumstances. That said, who's to say that something in the material makeup of everything from a tree to a block of granite (but of course, not a block of sand) could not allow it to become quantum locked in this world?

This is certainly the best theory I have so far, but there is one small problem with it. While the force keeping a quantum locked object levitating is extremely strong, it is not infinite. If you pile enough weight on top, it will eventually fall down to the surface. Now the amount of force, and the amount of weight it can therefore support, depends largely upon the types of materials being used (and which I know nothing about, given that we're talking about a tree quantum locking with a patch of dirt . . .).

There is still one experiment I can try, though. You see, any weight stack on top of a quantum locked object is transferred directly to the object beneath it, meaning that even if a quantum locked tree could support a hundred blocks of sand without moving, as soon as I stood under that tree, the entire weight of the tree and the sand would come down on me. A simple test showed that this does not happen here on Planet Minecraft (or at least doesn't seem to), when I went ahead and walked under the tree. That said, there is still the possibility that the density of everything is so low here that even a hundred blocks of sand would be only marginally heavy and not enough to cause any damage to my body. Also, I haven't really been able to find anything that can squish anything else simply by dropping it, so there's that too.

All that said, this does seem to be the best theory I have at this point. After all, even though I do not have a scale to confirm it, it is theoretically possible that the relatively low density and weight of the vast majority of materials here is responsible for them being able to maintain quantum locking (based on some unknown materials inside of the blocks and the ground). Furthermore, that would go a long way toward explaining why I am able to carry a few thousand blocks of granite in my pocket without breaking a sweat.

At the end of the day, though, any explanation that requires so many far-fetched possibilities is probably doomed to fail. For the time being, I will simply have to accept that gravity simply does not work here the way it does on Earth.

CHAPTER 5
PLANET MINECRAFT

O kay, I've been putting off discussing this particular aspect of my scientific explorations here, because I'm still having such a hard time wrapping my head around all of it. But I can't put it off any longer. Let's talk about Planet Minecraft.

PLANET MINECRAFT

Naturally, my first assumption when I arrived here was that I was on a planet much like any planet in our own solar system. I assumed it was a spherical ball of rocks and minerals, floating in space, orbiting around a star, being orbited itself by a moon, and that it was in a relatively fixed position to the stars in the sky. It did not take too long, though, for me to realize just how wrong (or at least problematic) these kinds of assumptions were. In fact, it took me precisely until the end of my second night in this world when, after finally feeling some sense of safety after being constantly pursued by monsters since my arrival, I looked to the east to watch my first Minecraft sunrise. That's when I saw something that instantly made me throw all those assumptions out the window. It was, of course, the shape of the sun—it's a square. A perfect, four-sided, right-angled square (like the moon, I would later learn).

This, of course, led me to the big question: Am I standing on a square planet?

Now I say square (as opposed to cube-shaped) because, from my vantage point here on the surface of this planet, I can't even be sure that any of the celestial bodies I can observe from here are fully three dimensional. After all, a cube only looks like a square (you know: four sides, four angles, one surface) when viewed straight-on. If you turn the cube even slightly in any direction, you will be able to see six sides, six angles, and at least two surfaces. When you look up from here at the sun, the stars, or the moon (at least when it's full) you only ever see a square. Still, for the purpose of the following explanation, I'm going to assume that the celestial bodies are, in fact, cubic, because the problems that arise from cubic planets and stars would mostly just be more extreme versions of the same problems with square planets.

Also: square planets? I mean, come on. I just can't even . . .

So, it's a natural question: Am I standing on a cube-shaped planet? The difficult part, of course, is figuring out how to answer it.

Well, first of all, let's get the obvious out of the way: No, unfortunately, I have not yet found a vantage point from which I can see far enough to the horizon to notice any curvature of the planet I am standing on. And believe me, I tried. I went to the ocean, I went to the desert, I even went to the top of a mountain and built the tallest tower I could to look out from, and I still couldn't see anything like the curvature of a planet (or, for that matter, the lack thereof). Unfortunately, while the air seems quite clear in my immediate surroundings most of the time, the atmosphere here is such that at a distance of over a few hundred meters, everything starts to get very hazy. So, even if there were a clearly flat or obviously curved horizon, I wouldn't be able to see it.

So, without being able to see the horizon (and not seeing a recipe in the crafting table for a rocket ship) how can I figure out the shape of the planet that I am currently standing on?

Now, I have to admit that the squareness of everything here at first made this seem slightly less strange than it should until I started to think about how a cube-shaped star could actually exist. To consider that, first, we have to look at how stars form in our own universe.

SUN AND STARS

All stars in our own universe, including our sun, begin as massive clouds of gas and dust. These clouds, known as molecular clouds, are made up of dust, atoms, and various molecules all just kind of swirling around out in space. Now, if you remember from our discussion of gravity, all mass in the universe has a gravitational attraction to all other mass in the universe. That's right, I said *all mass*—even the clump of mass that you call your own body exerts a gravitational attraction. If you clip your toenail, that toenail clipping has gravity. You blow your nose, your boogers have gravity . . . you get the picture. Of course, the amount of gravity created by a booger is extraordinarily small, and not nearly enough to attract anything that has any substance to it. Similarly, each molecule of dust floating around in the enormous cloud in space will have a tiny little amount of gravity, too.

Now, it's important to keep in mind that all of the dust and gas molecules that make up this molecular cloud are not standing still. In fact, they are constantly swirling around, turbulently moving from one place to another, and as a result, they are never evenly distributed throughout the cloud. When this constant motion continues for millions of years, sooner or later you will wind up with a region of the molecular cloud in which enough particles of matter have swirled close enough to each other that they begin to collapse due to the pull of their collective gravity.

Of course, this clump of gas and dust is much smaller than the cloud itself (though, for comparison, even a small clump like this would be much larger than our entire solar system) but as its mass grows, it attracts and accumulates more and more gas and dust. As it accumulates more gas and dust, its mass increases even more, which makes its gravity increase, which makes it attract even more gas and dust, until eventually, you wind up with what is called a pre-stellar core. This is basically when the growing clump of mass and dust becomes large enough to start to generate heat. Over time, more and more matter is pulled into the core, compressing the interior to greater and greater pressures, until eventually, the pressure becomes so great that it causes the nuclei of hydrogen atoms at its center to begin breaking apart in a chain reaction (also called a nuclear reaction). This nuclear reaction (or rather, the enormous amount of heat produced by the nuclear reaction) is what makes the star begin to shine.

At this point, however, the new star is still not quite spherical in shape yet. It is actually more of a swirling disk shape with a burning sphere at its center. Over time, gravity will continue to pull some of the innermost material in that swirling disk into the star. Meanwhile, the matter on the outside of the disk starts to clump together from its own gravity until these clumps get large enough to become planets. As the disk finally dissipates, the planets continue their motion around the star, which is how they wind up with the kind of orbiting motion we observe in our own solar system.

The important thing to remember here is that all of this—the entire process, from dust cloud to solar system—is caused by gravity. Whether you are talking about a star or a planet, the formation essentially occurs by clumps of material in space pulling more material in, growing in mass and gravity, and continuing to grow until the material runs out. Now, because gravity extends evenly in every direction, the basic shape will always be somewhat spherical. Sure, like Earth, most planets and stars are not perfect spheres, but this is due to the other forces acting on them and the variation of mass and density within the material that makes up the planet or star.

All of this is to say that there is no possible way, based on everything we have learned about our universe, that this process of star and planet formation could ever create a stellar body like a star in a cube shape, pyramid, trapezoid, or anything else. This is fairly simple. A sphere is the only three-dimensional shape where the surface is always roughly the same distance from the center no matter where on the surface you measure from. This makes perfect sense with gravity, as the gravity of a particular object is always exerting the same amount of gravitational force on everything around it, with the gravity decreasing at a constant rate the farther you move from the center of the object.

In a cube, on the other hand, the distance from its center to the middle of one face of the cube is considerably less than the distance from the center to one of its corners. This means that the force of gravity acting on whatever is at the surface in that corner would be far less than the force of gravity acting on the center of the face of the cube. So, with the laws of physics being what they are in our universe, even if you were to somehow magically materialize a cube-shaped planet in the middle of space, it wouldn't stay in its cubic shape for very long. Eventually, the corners of the cube, with their lower gravity, would either start to topple in toward the center of the faces or else simply fly out into space (depending on whether or not they are heavy enough to be attracted by the gravity at the center of the cube). Thus, even if you did start with a cube-shaped planet (which is of course impossible, I know, I'm just saying . . .) you would wind up with the sphere-shaped planet long before anyone showed up to start walking around on its surface. The same process, of course, would take place if you started with a flat, square-shaped planet.

So then, how is it that a cube-shaped star does exist, right here above my head? Well, I don't really have a good answer for that, except to refer back to the fact that gravity does not work the way it usually does on Earth here on Planet Minecraft. My only theory is that whatever it is about the function of gravity in this universe that allows a tree to remain locked in place after you remove the lower third of its trunk, must also be responsible for allowing a star to form in a cube or a square.

Yeah, I know that's not much of an answer, but it's the best I've got. And unfortunately, it still doesn't even begin to help us with the problem of the moon. And whoa boy, does the moon present some problems.

THE MOON

At first glance, you would think the moon is more or less the same as the sun. A cube (or a square) floating in the sky, orbiting a planet. Also, being that planets and moons form more or less the same way that stars do, it would be natural to surmise whatever crazy work of gravity allowed for a cube-shaped star, would also allow for a cube-shaped moon. This is what I thought for a little while anyway. That is until I began to notice something strange about the phases of the moon.

Here is what I found so strange about the phases of the moon that orbits Planet Minecraft: nothing. That's what's so strange.

You see, the moon here seems to follow more or less the same lunar phases that the moon on Earth does, albeit with a considerably shorter lunar cycle (on Earth, our moon phases have a twenty-nine-day cycle, while here the moon has an eight-day cycle). That's what is so strange: a cube-shaped moon would not have phases like our spherical moon does back on Earth. Not even close.

Allow me to explain.

You see, on Earth, the moon orbits around our planet once every twenty-seven days (yes, I know I just said it has a twenty-nine-day cycle, but it needs that extra two days to catch up to the Earth, which travels nearly 73 million kilometers [45 million miles] during each cycle). If you thought the moon orbited around the Earth once per day, and that's why it rises and sets every day, you would be mistaken (although no more mistaken than humans were before the Copernican Revolution). In truth, it only appears that the moon orbits every day because the Earth is also rotating at the same time the moon is orbiting. Let's set that aside for now though and focus on the orbit.

Due to the fact that the moon is constantly moving in a circle around the Earth, its angle toward the sun is constantly changing, even though its angle toward the Earth is always the same. So, when you see that the moon is full, that is just because the side of the moon that faces the Earth on that particular day is angled directly toward the sun. When the side of the moon that faces the Earth is angled directly away from the sun, you get what we call a new moon, where you can't really see the moon at all except as a dark shadow in the sky.

Are you with me so far? Good, because here's where things get weird.

You see, even if we did have a cube-shaped moon, or somebody simply built a huge cube of cheese and shot it up into the sky to orbit the Earth, you would still have a full cube moon whenever the entire side facing you is also facing the sun, and a new cube moon whenever the entire side facing away from you is facing the sun. What you would definitely not have, however, are any of the other phases of the moon: crescent, quarter, gibbous, or otherwise.

Think about it this way: If you stand in a dark room holding a soccer ball in your left hand and a flashlight in your right hand and point the flashlight directly from your nose to the soccer ball, you will see the entire circular shape of the ball. At the same time, you won't see any of the circular shape of the ball if you hold the flashlight directly on the opposite side of the ball. This much is obvious. But then think about what would happen if you were to slowly rotate the flashlight from the back side of the soccer ball to the front. Well, the first lighted shape would be a crescent, as the flashlight illuminates just a sliver of the side of the ball that is facing in your direction. As you continue to move the flashlight, the sliver of a crescent will get larger and larger until you are pointing the flashlight at the ball at a right angle to your own vantage point, at which time you will see exactly half of the ball. Keep moving the flashlight, and you will see a slightly more rounded version, though not quite fully circular (this is what we call a gibbous moon) until you finally wind up with your flashlight illuminating a full circle, and then the whole process continues in reverse.

Now, imagine trying that same process with a perfectly square cardboard box. What would happen? At first, with the flashlight pointed at the back of the box, you wouldn't see any shape illuminated by the flashlight. Then, as the flashlight makes its way to that right angle position, you still wouldn't see any light shining directly on the part of the moon facing you. From the point where you would have a new moon to the point where you would have a quarter moon, you do not see any light shining on the box at all. In fact, you won't see any light shining directly on the forward-facing side of the box until the flashlight moves past the quarter phase and starts to illuminate the front of the box directly. Even at that time, however, you wouldn't see a gibbous shape; you would see a straight line of light extending from the top of the box to the bottom of the box, fading into darkness as it extends across the surface. This straight line of light would get wider and wider as you moved the flashlight until it faced the box head-on, and you'd get your full moon (or full box, as it were).

Are you starting to see where I'm going with this? That's what should be happening with the square moon that you can see here from the surface of Planet Minecraft: it should be completely dark for the first quarter of its lunar cycle, followed by a quarter cycle in which a small, straight strip of light begins at one side and grows night by night until it illuminates the entire front face of the cube-shaped moon, a quarter in which the illuminated portion shrinks back down to a line of light from the top to the bottom on the opposite side, followed by a final quarter in which the moon is completely dark once again. But that's not actually what happens here. Not even close.

Here are the actual phases of the moon in Minecraft, as they look when watching the moon shortly after it rises over the eastern horizon:

New Moon: The new moon is dark, with a slight square outline of light around its edges. This looks very much how you would imagine a cube moon should look if the sun were directly behind it.

Waxing Crescent: The top left corner, left side, and bottom left corner are illuminated, with a crescent-shaped *arch* of shadow cutting into the left half of the moon, and the entire right side in darkness. It sort of looks like the shape of a square cracker if you took a big old bite out of it. Considering that I am describing this while looking east, it appears that the light is coming from the north.

Waxing Half: With the waxing half moon, exactly one half of the moon is illuminated, so it looks like a rectangle half the size of the full cube moon, though it is not illuminated evenly, nor is it illuminated with a straight line of fading light, as the box in our imaginary experiment would be. Rather, there is a semicircle of bright illumination on the north (or left, from my vantage point) side of the rectangle, with the remaining surface of the rectangle on the north side slightly less illuminated, but still visible. The south (or right) side of the moon is completely dark. You could possibly replicate something similar to this with the box and the flashlight if you held the flashlight very close to the box (as in, between your eyes and the box) but even then, some light would spill onto the other half.

Waxing Gibbous: Perhaps the strangest shape of all is the one that occurs during the waxing gibbous phase. With this phase, the left half of the square is illuminated fully, while the right half of the square shows only a more dimly lit circle shape. Imagine the shape you would have if you began with a square piece of paper, drew a circle that touched each of the four sides, and then only cut out half of that circle. Also, the bottom of the moon (again, as seen over the eastern horizon during moonrise; it would be the top of the moon as it sets in the west) is considerably less illuminated than the top. For the life of me, I can't imagine how you would replicate this with a flashlight and a box.

Full: With the full moon, you can see the entire square, though it is important to note that the left and bottom edge of the square are slightly less illuminated than the rest of it.

Thankfully, after this, the phase of the moon follows a predictably reverse waning cycle, with each phase being more or less a mirror image of its waxing counterpart.

So now the big question: How in the mother living heck is any of this possible?

Obviously, the phases of the moon here in Minecraft are not being caused by the same set of circumstances that cause the phases of the moon on Earth. There is simply no way that anything akin to the motion of the Earth, moon, and sun in our solar system could cause these specific shapes to occur in this specific order, no matter what shapes any of the participating celestial bodies happen to be.

So, if the phases of Planet Minecraft's moon are not caused by the same circumstances that cause the phases of Earth's moon, what is causing them?

My first thought, upon noticing the shape of the waxing crescent moon, was that I was simply seeing the shadow of the Minecraft planet being cast upon the moon. Assuming that the planet I am standing on here in Minecraft is in fact a sphere (more on this later), this shape could be caused simply by the sun being set slightly off-center, causing most of its light to get blocked by the planet itself. This is essentially what happens on Earth when we have a partial lunar eclipse, and would at least explain how a cube-shaped sun causes a rounded crescent-shaped shadow to appear on a cube-shaped moon.

Unfortunately, my lunar eclipse theory completely breaks down as soon as you see the next phase of the moon: the waxing half. Remember, the overall shape of the waxing half is a rectangle exactly half the size of the whole moon, but with the left (north-facing) side of the rectangle more brightly lit than the right (south-facing) side, and with the more brightly lit north-facing side having a decidedly rounded border. If my assumption about the waxing crescent was correct, and the shape of the moon phases was caused by the Minecraft planet coming between its moon and its sun, then whatever motion caused the sun's light to go from being completely blocked to only partially blocked would have caused this next phase to simply look like a larger crescent, with a smaller semi-circle-shaped cut out on the right/south side of the moon and a larger portion of the left/north side of the moon illuminated.

Again, as I alluded to with the flashlight and the box, the only way I can even imagine a shape like the actual waxing half occurring naturally would be for a spherical sun to be very small and very close to the moon, and with the square moon turned just slightly in the sun's direction. Of course, the sun here is clearly not round, and while I can't tell how far away the sun is, it certainly doesn't seem close enough to create a curved light on the moon!

That said, if the sun were in fact small and round and very close, the next phase, the waxing gibbous, would be a plausible progression. In fact, now that I think of it, the small, close, round sun theory would be slightly more plausible than the lunar eclipse theory for pretty much all of the phases of the moon here . . . except for the fact that the sun doesn't seem nearly close enough and it definitely isn't round.

This leaves only one possibility that I can think of: The moon is not, in fact, being illuminated by the sun at all. Rather, it must be getting its illumination from some other light source altogether (albeit one that I cannot see from the surface of the planet). I know, it seems improbable, but really, what else could it be? There is simply no way that a square light source like the Minecraft sun could create the strange and unique shapes of the moon's actual phases.

This theory is also bolstered by the fact that the sun and moon seem to remain exactly 180 degrees from the Minecraft planet at all times and with no observable changes in either the time or angle of their respective rising and setting times and locations.

If there were some other light source that was simply not visible to me from my region of the Minecraft planet, *and* this light source was relatively small (compared to the moon at least), *and* the light source itself moved back and forth in relation to the position of the moon . . . Well, then, theoretically that would explain it.

This may also help to explain why the stars don't move. That's right, the stars don't move here. Well, that is to say, they don't move relative to the sun and moon, at least.

This is obviously very different than the way that the stars behave in the sky on Earth. Just go outside on any night, look up at the sky, and take a look at the constellations of stars immediately adjacent to the moon, you will find them changing constantly—both throughout the course of the night and from one night to the next. In the same fashion, if you look at the place on the horizon where the moon rises versus where the sun rises, you will also find them constantly changing. This occurs on Earth because our planet is constantly rotating while the moon is constantly orbiting the Earth, and the Earth is constantly orbiting the sun. With each of these three bodies constantly changing position along so many vectors, the relative positions of the sun and moon to any spot on the surface of the Earth changes from day to day. However, as I explained above, none of those circumstances explain what you can actually observe happening here in the Minecraft world. If you stand in the same spot without moving for a few days, you will see the moon will rise over the same exact point on the horizon every evening, follow the same straight line across the sky, and set in the same place every morning, at which time the sun will peek over the same point on the horizon that the moon rose the previous evening (and yes, it follows the exact same path across the sky and sets in the same place too).

Remember, though—we're not talking about the sun and the moon anymore; we're talking about *the stars*. That's the really strange part: the stars follow this exact same path as the moon and the sun. You can pick out any constellation of stars in the sky (I tend to use the constellations near the moon because they are easier to find, but it works with all of them) and watch where it rises and sets every day, and it will not change. Not ever.

So what does this mean, exactly? Well, for one thing, it means that the moon here must not actually be orbiting this planet at all, and this planet (or whatever it is) isn't orbiting the sun either.

Yeah, I know how crazy that sounds, but stick with me.

Think about it this way: Assuming the stars I see in the sky here are actual stars, and not just little bits of lint floating in the sky (or whatever), they have to be very, very far away. Now, while I can't calculate exactly how far the moon is from the surface of this planet, I can certainly assume that it is much closer than the stars. This is the problem: if something that is very close to you (and which appears to be moving, at least in relation to you) does not move in relation to something much farther away behind it, the only conclusion to be made is that neither object can be moving at all. Add to that the fact that the sun also never moves in relation to the moon (it's always exactly 180 degrees opposite the moon), and the only logical conclusion you can draw is that the sun, the moon, and the stars are all more or less stationary, and the only thing causing the cycle of day and night and the apparent motion of these celestial bodies is the rotation of the planet I am standing on right now.

So then, what does this mean? What does it say about the physics here compared to the physics back on Earth? Well, it says a lot. But before we get into it, we have to pick up where we left off in the previous chapter on the story of Albert Einstein and his Theory of Relativity.

As I explained in the discussion of gravity from the previous chapter, after discovering that gravity and acceleration were functionally identical, Einstein set out to discover *why* they were identical. What he figured out was as mind-boggling as it was brilliant: space and time are not actually separate things at all, but one thing called spacetime, and every ounce of mass in the universe makes this thing called spacetime curve.

Confused enough yet? Yeah, I thought so. Let's do another thought experiment to make these ideas a bit easier to understand.

Imagine you take a large sheet of thin rubber, maybe 10 meters (33 feet) in diameter, and attach it to a metal frame which you then suspend a few meters off the ground. You then take a solid steel cannonball and place it in the center of the sheet. The weight of the cannonball causes the rubber to stretch toward the ground, creating a funnel shape out of the sheet. Now let's say that you also have a tiny marble, just a small fraction of the size and weight of the cannonball. What would happen if you were to drop the marble onto the edge of the sheet? Obviously, it will simply roll straight down to the center of the funnel until it falls into the cannonball at the bottom.

This is basically how gravity works in Einstein's Theory of Relativity. The sheet in this experiment is spacetime, and the mass of the cannonball is actually bending it in such a way that a smaller object (like a marble, for example) will have no choice but to fall into the larger object. It's the same as if I were to build a tower from the surface of the Earth all the way to space and then jump off it, I would fall straight down to the ground, more or less (not counting variations due to air resistance, wind, or the rotation of the planet).

Now, imagine what would happen if you took that same marble, and instead of dropping it straight on the edge of the rubber sheet, you pushed it in a sideways direction, perpendicular to the direction of the cannonball.

What would happen then? Well, the marble would first start to circle around and around and around and around the sheet until it finally lost momentum and fell straight into the cannonball. The speed is the key here: if you just give the marble a little push, it won't go around too many times before falling into the cannonball, but if you were able to push the marble at a very high rate of speed, it would circle for a very long time before finally falling into the center.

This, essentially, is what happens when an object orbits a planet, or a planet orbits a star. It is simply the natural motion of something that is being pulled by gravity toward the object causing that gravity while at the same time carrying momentum perpendicular to that force of gravity. Now I know what you're thinking: If that's true, why hasn't the Earth fallen into the sun? Or why hasn't the moon fallen into the Earth? And the answer to that is simple: eventually, if nothing else changes, they will. Without a force being exerted on an object to keep it in a stable orbit (for example, something like a space station has engines that burn fuel to keep pushing it in the right direction and maintaining its altitude) every orbit will eventually decay until the object falls into the mass that is exerting the force of gravity upon it.

In other words, the faster a moon is orbiting its planet, the longer it will take for that moon to fall into the planet it is orbiting. This means that since the moon here on Planet Minecraft does not appear to be orbiting at all, it should have fallen into the planet a long time before I ever got here. Of course, the same idea applies to planets and stars as well. The Earth orbits the sun at approximately 30 kilometers (19 miles) per second. If you were to invent a magical handbrake and slow that down to zero in an instant, the Earth would simply start to plunge directly toward the sun until it burned up in the star's Corona.

And yet, after many, many days (or months, or years . . . it's so hard to keep track of that sort of thing when the days are so short) here on Planet Minecraft, neither the sun nor the moon has crashed into the surface of the planet that I'm standing on. But what could possibly explain this?

Once again, the obvious answer is not very satisfying: something is seriously messed up with gravity here.

That said, until I have a better theory for what exactly is happening with the gravity on this planet, I should at least entertain other possibilities for how to explain these strange phenomena. The only theory I can think of at the moment is still rather far-fetched, to say the least. Here it is: All those lights in the sky—both the little ones that seem to be really far away and the big one that seems to be really close—are not actually stars at all. If they are some other kind of ball of light and heat (and the ones that look like stars may be altogether different than the one that looks like the sun) and they aren't actually very far away at all, then it is theoretically possible that everything I see in the sky is in fact orbiting around the planet that I'm standing on after all.

Imagine, if you can, a planet too far away from any stars to actually see them, surrounded by comets, a moon, and some unknown object that is neither a comet nor a moon (and just so happens to look like a cube-shaped version of the Earth's sun . . .), and all of them are orbiting in such a way as to remain precisely in the same positions relative to each other. On such a planet, it would be theoretically possible to find the circumstances observable from Planet Minecraft without completely getting rid of everything we have ever learned about gravity on Earth. Of course, this is such a ludicrously improbable idea that I can hardly take it too seriously, but considering that the alternative is to throw my arms up and go with the "I Guess Gravity is Just Funky Sometimes Theory," it's probably the best theory that I'm going to come up with.

CHAPTER 6
MINECRAFT MUSHROOMS

O bviously, it goes without saying that the most striking and unique feature of the mushrooms on Planet Minecraft is their size. We have nothing quite like them on Earth, which is not to say that we do not have some very large fungi on Earth. For example, the *Armillaria ostoyae* is a type of fungus native to the Pacific Northwest of the United States, and it's not only the largest fungus on the Earth, it may very well be the largest living organism on the face of our home planet. In fact, one particular organism of this fungus in Oregon covers an area of more than two thousand acres and has been estimated to be nearly seven thousand years old.

Now, before you go picturing a single mushroom the size of a small city, note that I said *the organism* is larger than two thousand acres, not the mushroom. You see, like many fungi, this particular species creates many mushrooms that rise up out of the ground, or form on trees, while the bulk of the organism itself exists as mycelium underground.

This is because, on Earth at least, the mushroom is only the fruiting body of the fungus organism, while the main part of the organism is made up of underground mycelium. The important distinction is that, as a fruiting body, the mushroom will go through its growth, death, and decay cycle while the organism itself continues on beneath the soil. In this way, it is very much like an apple on an apple tree. You don't think of the apple as the plant; you think of the apple merely as a product of the

66

plant. If you pick the apple, the plant will go on living and just produce more apples. The purpose of the apple is to spread the seeds of the apple tree. Likewise, with fungi, the purpose of the mushroom is to spread the fungi's spores. Spores do function very similarly to seeds; they can be blown by the wind or spread by animals in order to create more versions of the fungus from which they originated. The primary difference between spores and seeds is that spores consist of only a single cell that is capable of growing into a new multicellular fungus, while seeds are themselves multicellular embryos of their parental plant.

When you look at the way many mushrooms work on Earth, they can appear at first like trees, with roots coming out from the bottom spreading into the soil. While the appearance is very similar to that of a tree's roots, the function is actually quite different. With mushrooms, the stringlike strands that spread out underneath the surface of the soil are not merely roots that feed the organism above, but rather they are the organism itself. That is to say, they are the part of the organism that continues living while the mushroom it produces grows, spreads its spores, and then dies, allowing the organism beneath the surface to grow another mushroom (or, in some cases, many mushrooms). The part of the organism beneath the surface that looks like roots is called the mycelium. With some species, the mycelium of a fungus will grow a single mushroom, while other species will grow many mushrooms over a large area, like the above-mentioned *Armillaria ostoyae.*

Now, while only the enormous fungi living on the Earth today have nearly all of their mass contained in their mycelial network beneath the soil, there were, in prehistoric times, some species of fungi that created mushroomlike structures that were in fact quite tall. Maybe not as large as the mushrooms here in the Minecraft world, but still quite larger than anything we see on Earth today. The largest of these fungi is known as *Prototaxites,* which lived on earth somewhere around 400 million years ago. Although we do not know a great deal about this species, we have found fossils of this type of fungus with a diameter of one meter (3 feet) and a height of nearly nine meters (30 feet)—quite similar to the size of the mushrooms here on Planet Minecraft. It should be noted, however, that these particular types of fungi did not have a traditional mushroom cap spreading out from the top of their trunk, but rather consisted only of a trunk, with occasionally a few branching trunks coming off the main trunk.

Still, as unique and wild as these large mushrooms here in the Minecraft world appeared, they did not seem to me on first consideration to be anything out of the realm of possibility on Earth. That is, until I started to actually inspect them more closely.

As mentioned above, mushrooms on Earth are merely the fruiting body of the organism beneath the soil, which means that if you dig beneath a mushroom on Earth, you'll usually find its network of mycelium moving out into the soil beneath it. However, when I ventured out to do some research on the mushrooms here and dug beneath these behemoths, I didn't find anything like mycelium at all. As far as I could tell, they didn't seem to extend into the soil at all.

Of course, by this time I had also realized that trees in this world do not have roots, so I can't say I was terribly surprised by this fact, but as noted above this is a very different thing when it comes to mushrooms. Finding a mushroom without mycelium would be more like seeing an apple growing on the ground and not finding a tree attached to it anywhere.

It was quite a long time after my first attempt researching the giant mushrooms of Planet Minecraft that I made a discovery that truly changed everything I thought I knew about fungi.

Shortly after I built my first boat here in Minecraft, I ventured far out into the western sea in search of new biomes, materials, animals, etc. My journey was, at first, a bit disappointing, as I mostly just found a bunch of sandy islands covered with the same types of vegetation I had seen in the forested lands around my home. Unsatisfied with my lackluster findings, I decided to keep venturing until I couldn't see land anywhere around me anymore. That's when I saw it: off in the distance, an island unlike anything I had ever come across before on Earth or on Planet Minecraft. It was a mushroom island.

At first, I just thought it was an island with a particularly large number of giant mushrooms towering on the shore. But as I finally pulled my boat ashore I found something that tipped me off right away that this place was different—the soil. The soil on this island was unlike anything I had seen anywhere else on Planet Minecraft so far. It had a creamy, purplish coating on the top of it, and tiny black dots kept appearing over the surface and then floating away. I tried to catch a few of these little dots, but they were either too small or too light and I could not get ahold of any before they disappeared in the wind. It was then that I looked up and noticed that it was not just the shore that was covered with mushrooms, but rather an entire island with literally hundreds of the enormous fungi, including some that were taller than any I'd seen in the forest. That's when I realized what the strange coating on the island's soil really was: the mycelium. Of course! How could I be so dense? Obviously, these mushrooms needed some form of mycelium to come from, and this must be it. Like the huge fungus in the Pacific Northwest, this was clearly an entire island covered by a single continuous organism of mycelium-sprouting mushrooms everywhere.

Of course, there were a few things that didn't quite make sense, at least in terms of the way that mushrooms work on Earth. First of all, both the red-capped mushrooms and the flat brown mushrooms seemed to be sprouting out of the same mycelium. The only way that this could happen would be if they were actually part of the same organism, and therefore the same species. I had certainly never heard of any Earth fungi growing different fruiting bodies that were this disparate in appearance (or, for that matter, any organism on Earth growing fruiting bodies that look so different) but out of all of the crazy things I had already accepted in this world, that didn't seem too far out there.

The big question for me was this: How was it possible for mushrooms to sprout so far away from this mycelium? If, like on Earth, it is the mycelium that is the actual organism, then certainly only mushrooms that were connected to it could grow. I pondered this as I walked around the island until I came across something so strange, it made me think that perhaps the mycelial network of this organism contains powerful mind-altering chemicals (as some fungi on Earth certainly do). What I saw was a cow, but not like any cow that I had ever seen on Earth, this world, or even in my wildest imaginings. This was a red cow with white spots and mushrooms—yes, mushrooms—growing out of its back and head. I looked around and noticed that there were literally dozens of these fantastical creatures wandering the island.

These mooshrooms, as I would call them, did not, in fact, turn out to be a hallucination at all. And while I cannot say that I have a very clear understanding of how these organisms came to exist, they did give me a theory about how it was possible for these mushrooms to grow so far away from this mycelial island. My best guess for how such a strange-looking cow could exist is that the spores from this species of fungus simply must be some of the most efficient spores in existence, or at least far more efficient at their job than any of their counterparts on Earth. My thinking is that the spores must travel on the wind from these mushroom islands and then land in the forests and swamps of the mainland. There, they must be able to sprout up their fruiting mushrooms with only the tiniest bit of mycelium beneath them. The reason why this seemed to make sense to me at the time was that only an organism so efficient at reproducing itself could manage to grow, not only on various and wildly different kinds of soil, but even on the backs of cows (and maybe even more animals, for all I knew).

All that said, as strange as these creatures looked, having fungus grow on a mammal is hardly unusual to our own planet. Just walk into any pharmacy and you will see an entire shelf filled with antifungal creams designed specifically to kill the various types of annoying fungi that grow on humans. While athlete's foot, or other similar fungi, do not change your entire skin tone, they certainly can change the color of the skin directly beneath where they are growing. And this simply must be what's going on with these cows. Somehow, they made their way to this island and found that this mycelial growth on the soil, or maybe even the mushrooms themselves, are quite good food for them to eat, but as a result of staying here so long they wound up absolutely covered with this fungus. Of course, to confirm this theory I would have to find a newborn mooshroom, but the possibility of winding up with red skin and covered with mushrooms myself convinced me not to remain on the island long enough to observe a live birth. . . .

CHAPTER 7

MINECRAFT BUGS

This seems like as good a time as any to talk about bugs—this world's wide variety of creepy crawlies. And oh boy, are the crawlies here creepy! Like most humans from Earth, I have a natural fear (or at least a distaste for) most bugs, and spiders especially. Yes, yes, I know. As a scientist, I am well aware that Earth's complex ecosystem would not function without these magnificent creatures, the vast majority of which are completely harmless to humans and certainly never did anything to me personally. Still, I can't help it. Bugs have just always grossed me out. And the bugs here in Minecraft world, being far more massive than anything on Earth, gross me out far more than any bugs on Earth.

Nonetheless, my mission here is to explore, research, and communicate my findings, and that's what I'm going to do . . . no matter how gross it gets.

In all probability, the first creepy crawlie that you are likely to encounter if you are ever fortunate (or unfortunate, as the case may be) enough to visit this planet is the Minecraft spider. It's pretty hard to miss them, because like a lot of things here, they are freaking huge. As you may recall from my description of my first night here in this world, the spider was one of the first living creatures I encountered, and one of the reasons I ran so far and for so long before stopping to catch my breath.

How big are they exactly? Well, based on my calculations, these terrifying creatures are about two meters (7 feet) wide, two meters (7 feet) long, and almost a meter (3 feet) tall. Have you ever seen a Great Dane? You know, those huge dogs that look more like small horses than canines? Now imagine a spider twice the size of one of those enormous animals.

Like I said: absolutely terrifying.

71

Perhaps just as bad as the size of these arachnids is their charcoal black bodies and glowing red eyes. (I never got a really good count of exactly how many eyes they have, but I've counted at least a dozen.) For reference, the largest spider on earth is the Goliath Birdeater, a member of the tarantula family that lives in South America and can grow up to 13 centimeters (5 inches). While this particular species of spider is not actually dangerous to humans (if they bite you, the effects of their venom are comparable to a wasp sting), if you were to actually see one of these behemoths in person (which I have) you would probably not want to get close enough to find out. Furthermore, while the history of Earth does contain some extraordinarily large creatures of all shapes and sizes, nowhere has anyone ever found evidence of a spider as large as the ones you see here in the Minecraft world.

Now to be fair, there is a second species of spider here (the cave spider) that is both rarer and considerably smaller than the one referenced above. Even these relatively diminutive cave spiders, which are roughly half the size of their topside dwelling relatives, are far larger than anything that has ever lived on Earth. That said, it should also be noted that, as with many arachnids on Earth, the smaller variety is actually the deadlier one. When the cave spider bites you it transfers a toxic venom that can put you out of commission for quite a long time, while its larger cousin relies merely on biting you with its enormous fangs to infect damage (though this is not exactly comforting, I know).

Now, while we may not have ever had any arachnids as large as the spiders here in Minecraft on Earth, it's important to remember that arachnids are merely one type of arthropods, many of which (especially in prehistoric times) can grow to be far larger than any spider. The largest living arthropod today is the Japanese spider crab. This gargantuan crustacean can grow to nearly four meters in length from the tip of one claw to the tip of its opposite claw. That said, the actual body of these crabs is much smaller than this, rarely measuring more than 35 centimeters (14 inches) in width—not that this makes them any less the stuff of nightmares. Another similar giant crustacean is known as the coconut crab, which resides on many islands of the Pacific and Indian oceans. While the coconut crab does not measure as long from claw to claw as the Japanese spider crab (coming in at a comparably small 1 meter in length) its body is actually the larger of the two, topping out at a terrifying 40 centimeters (16 inches). Also, it should be noted that, unlike the ocean-dwelling Japanese spider crab, the coconut crab lives on land and is thus far more likely to run across an actual human being in the course of its day.

So yeah, we have nothing as big as the Minecraft spiders on Earth. Not even close. Still, we must bear in mind that this world is clearly at a different stage of its evolution than our own, so we have to look all the way back into our fossil record to see if there ever was anything comparable to their size on Earth. In fact, the largest arthropod ever found to have lived on Earth is the now extinct *Jaekelopterus*. Looking more like a sea scorpion than a spider, this water-dwelling arthropod (which has thankfully been extinct for nearly 400 million years) could grow to more than two and a half meters in length, making it even larger than the spiders here in the Minecraft world.

Ironically, the giant *Jaekelopterus* of Earth actually does look quite a lot like a creature that can be found here on this Planet Minecraft; a creature that, in my personal opinion, may be even more terrifying than the spiders. The creature I'm talking about is the deadly silverfish. On Earth, silverfish are a fairly common household insect, though incredibly small in comparison to the silverfish here. Generally measuring in at about 1.5 to 2.5 cm, they are nowhere near the size of their Minecraft counterparts, which I estimate to be almost one full meter in length and about half a meter in width.

So, you know . . . roughly forty times the size of our terrestrial silverfish.

It is not merely their size that makes the silverfish here so frightening, though. On Earth, the silverfish is a small insect that you only really need to be afraid of if you happen to be an old book or a forgotten rug sitting in an attic somewhere, as their diet consists mostly of polysaccharides, which they often get from the starches that make up the adhesives of books and carpeting (but can also be found in things like coffee, dandruff, and hair, which they also like to eat). These Earth silverfish, however, certainly do not ever attack human beings. On the other hand, not only do the silverfish in this world attack humans, if you have the audacity to try to fight back when they attack, they will call out to their friends and before you know it you are surrounded by dozens of these squirmy insects, all coming at you at the same time. As a result, I highly suggest avoiding this world's silverfish whenever it's humanly possible to do so.

The real question about all of these creepy crawlies for me, though, is this: What the heck makes them all grow so large? Not only are they all so much larger than their Earth-dwelling counterparts, as far as I've seen there don't seem to be any small (by any definition of that word) bugs here at all. Well, all I can do to answer that question is to discuss some of the theories as to why these types of creatures used to be so large on Earth.

First of all, we actually cannot say for certain why so many organisms on Earth seemed to grow so large in prehistoric times compared to modern times (not only arthropods, of course, but dinosaurs, woolly mammoths, and many other types of species were far larger in Earth's distant past). For a long time, a popular theory among Earth's scientists was known as Cope's Rule. This idea, named after paleontologist Edward Drinker Cope, states that evolution naturally tends to favor organisms growing larger in body size over time. When mass extinctions happen, as happened on Earth when the dinosaurs were killed off, for example, the only living creatures that remain tend to be very small. These tiny little apocalypse survivors, now free from all of the larger predators that used to hunt them, will then begin the process of evolving larger and larger over time until another mass extinction comes to wipe them out as well. While there is certainly some evidence to back up this theory, at least in regard to certain species (particularly reptiles, for example) there are also many species that we know from the fossil record actually got smaller over time. Thus, while Cope's Rule may play a contributing role in the evolution of such giants, it certainly is not the only (or even primary) factor that can account for their great size.

Interestingly, with a few exceptions (like the axolotl, for example) the only creatures in the Minecraft world to grow significantly larger than their terrestrial counterparts are all arthropods. This is interesting because, when it comes to Earth arthropods, there is actually a widely held theory about why they used to be large in the distant past that may apply quite well to these enormous Minecraft creepy crawlies.

You see, arthropods do not have lungs, and thus they do not take in or process oxygen the way that mammals do, for example. While ocean-dwelling arthropods use gills to take in oxygen (not unlike most species of fish) land-dwelling arthropods take in oxygen directly through holes in their body called trachea. These tracheae are basically tiny tubes that bring the air from the surrounding environment directly into the arthropod's body, and then diffuse the oxygen into the body's cells through a network of air tubes. One of the reasons most scientists believe that insects are so small today is that, with oxygen making up only about 20 percent of Earth's air, the amount of oxygen needed for them to grow any larger would be too great to pass through the tiny tubes that flow through their bodies. However, for large periods during prehistoric times, there was actually quite a lot more oxygen in the atmosphere (peaking at more than 35 percent about 300 million years ago). This higher concentration of atmospheric oxygen provided Earth's many species of arthropods with the ability to absorb more oxygen through the same size tubes, allowing them to grow larger in size than they can today. While I do not, alas, have any way of testing the atmospheric oxygen levels here in this world, it is certainly a reasonable hypothesis to assume that the great size of these creatures may be due to the difference in atmospheric gases on this planet. The fact that we don't see, for example, chickens here growing proportionally in size the way that the arthropods do seems to support this theory as well. If I do run across any elephant-sized chickens, though, I may have to revisit the issue.

CHAPTER 8
MINECRAFT ZOMBIES

As I mentioned in my prologue, my first real run-in with what you could call civilization on this planet was on my second night in the Minecraft world, when I came upon the ruins of a village that had been overrun by zombies. Now, being the type of person who generally does not like to have my brains eaten, I hightailed it out of there pretty quickly, only returning in the daytime to scrounge for supplies (which, if you recall, is how I discovered a crafting table and kicked off my real journey of scientific exploration). And honestly, once I had successfully looted everything I could from the village, I never really gave it too much more thought. I had built my cave dwelling far enough away so that the zombies that seemed to gather there didn't spill over too much into my neighborhood, and I went on about my business of exploring other parts of this world.

Of course, I was certainly curious as to how the decrepit village got there in the first place, but I just assumed that whatever civilization or species or interdimensional travelers (or whatever) had created my crafting table and all of its amazing abilities had also built the village at some point, before either being scared away or killed by the vast amount of dangerous creatures that wander this planet by night. Furthermore, the state of decay that the village was in, alongside the fact that for some time afterward I never came across any signs of civilization that did not appear in a similar state of decrepitude, led me to believe that whatever went wrong with that village had happened a very long time ago. It could have been there for hundreds of years, for all I knew. Maybe even thousands.

As it turned out, I could not have been more wrong.

Sometime later, I was exploring the far west reaches of the forest I first found myself in here (where I now reside in my cliff dwelling, just beyond the eastern edge) when, through a thick layer of trees, I spotted some structures that looked like another village very similar to the one I had first found way back on my second night.

Thankfully, it was daytime, so I didn't worry about this particular set of village ruins being overrun by zombies at the moment, and I went to explore to see if there were any good chests to loot. You can imagine my surprise then, when I strolled up to what I assumed would be a long-abandoned ruin only to find a populated, seemingly thriving village, bustling with people who appeared more or less human. In a complete and utter daze, I wandered into the center of town, watching as the inhabitants of the village went about their day: growing crops, tending to their livestock, strolling through the village square.

Could this actually be real? I wondered.

To be perfectly honest, there were a few minutes there when I actually thought that I might be imagining the whole thing. I even went so far as to seriously consider the question of how I would even be able to tell for certain if I was having some kind of a vast hallucination or not. I mean, sure, the loneliness had really started to get to me by that point, but what really had me questioning my sanity was how the people in the village didn't seem to notice my presence there at all. They just went about their business as if I were completely invisible, or at least such an ordinary part of the village's life as to not be worth a second glance.

Before too long, however, a bald man with bright green eyes and a black apron walked out of what I assumed was a shop of some sort just on the edge of the village square and just stood there staring at me.

So what could I do? I waved and said hello.

Of course, the man did not respond. I say "of course" only because even if these people could have anything like real conversations (which, it turns out, they can't), it probably wouldn't be in any language I could understand. In any case, the man just kept staring at me as I tried in vain to start a conversation until finally he took a small book out of his pocket and opened it up for me to see. Thankfully, the book contained pictures instead of words, and I was happy to see that they seemed to be the same types of pictures I found in the crafting table.

It didn't take me long to figure out what the pictures represented, and why he was showing them to me.

From what I could gather, the man was a stonemason and the book was a list of prices he charged for various stone items. I tried to tell him that I didn't need anything made out of stone at the present moment, but it quickly became quite clear that he was only interested in communicating with me about the items in the book—which stone items he had on offer (bricks and carved blocks of stone), and what types of metals and jewels he wanted in exchange for them (iron ingots and emeralds). Thinking that perhaps he was just a particularly shy mason, I bid him farewell and walked off to see if I could find somebody else to talk to. Alas, after trying to start conversations with pretty much every person in that village, I eventually came to the conclusion that, while they were perfectly happy to communicate about the various items they were willing to trade, they lacked either the ability or the desire to communicate about anything else. While I did wonder if perhaps they just didn't like strangers (which, on a planet where you are constantly attacked by zombies, giant spiders, and exploding monsters, I certainly had no trouble understanding), it didn't change the fact that they weren't talking to me.

Eventually, I reasoned that my next best step would probably be to go out and find some of the items that they wanted to trade for in the hope that trading with them would open them up to further communication. By that point, however, it started to get dark.

That's when I thought again about that first village I'd found and remembered that it seemed to be an absolute magnet for all the zombies, skeletons, and other violent creatures of the night.

Thankfully, the one thing I did have with me was my sword, so I took shelter in what seemed to be an empty house and decided to observe what would happen during the night hours to see if I could learn anything more about these people.

Night fell, a waxing moon rose over the eastern horizon, and within seconds I heard the familiar moan of a zombie just outside the door of the house. Steeling myself for a fight, I stepped out into the village square to see how these villagers would handle the invasion. I say invasion because that's how it was in that first village I found. As soon as night fell, the entire place was completely overrun with zombies and other monsters, so naturally, I assumed this would be the case in this village as well.

I quickly saw, however, that this was not going to be the case in this new village. When I first stepped outside there was just the one zombie and while more did arrive throughout the night, there were nowhere near the numbers that I experienced in the abandoned village. There's another thing that puzzled me as well: These zombies seem to look different than the other ones. Or rather, these ones all seemed to look the same, whereas I distinctly remembered the zombies in the abandoned village being more unique in appearance.

In any case, I clearly needed to learn more about the zombies on this planet, so I set off back to my home to find some of the items the villagers wanted to trade for, and to give that other village a good looking over.

When I finally made it back across the forest, I found the abandoned village much as the last time I had seen it: totally decrepit and empty. I found a house near the center of town that had not been completely destroyed and took some time to rebuild it. Once I finished my work, I went inside and waited for the night. This time, I can assure you that I did not venture out into the village square, because within a few minutes of night falling the town was once again completely overrun. Zombies seemed to come from everywhere here, not just wandering out of the forest at random. Before too long, though, the horde of zombies seemed to sense my presence, and they surrounded the house. At first they just kind of wandered around the outside of the structure, but they soon began to gather at the front, moaning even louder and banging on the door, trying to get in. Just when I thought I had really made a terrible mistake, I saw exactly what I'd come there to see: different-looking zombies.

Peering out the window to get a better view, I realized that these zombies were not only different looking than the other ones: they looked very similar to the villagers I had just met on the other side of the forest.

Now, when I originally started thinking of these creatures as zombies, it was really just because they were moaning and walking with stiff legs like something out of one of the zombie movies I had seen on Earth. But that night, as I saw what were clearly zombified versions of otherwise healthy humanlike beings, I realized that perhaps these didn't just look like zombies—perhaps they really were zombies.

Now, I know what you're thinking: zombies are fiction; zombies aren't real; as a scientist, you can't possibly believe that zombies actually exist, even if it is on another planet or in another universe (or whatever this place is). Sure, that's what I would have thought, too. But let's just take a look at the science here, shall we?

ZOMBIES ON EARTH

The first report of zombies on Earth came from the Caribbean island nation of Haiti in the early twentieth century. Now, of course, these weren't actual zombies (as in, people who died and then were somehow resurrected from the dead, filled with an unquenchable appetite for brains). There were, however, some seriously messed-up charlatans posing as sorcerers (known as *bokor* to the native Haitians). One of the powers that these "sorcerers" would claim to possess was the ability to bring the dead back to life. Of course, this is not the kind of claim you can make too often without proving it, but these bokor had a way to do just that.

The way that the bokor accomplished such a seemingly impossible feat began by combining, in addition to many other mysterious ingredients, the seeds of a *Datura stramonium* plant with the venom of a pufferfish. The venom of a pufferfish contains a potent neurotoxin called tetrodotoxin that, when ingested, blocks the sodium channels inside the brain, which causes the nervous system of the poor sucker who ingested it to lose its ability to send messages between the brain and the muscles, thus rendering the victim essentially paralyzed. *Datura stramonium*, meanwhile, is a powerful (and powerfully unpleasant, by all reports) hallucinogen, with the ability to put anyone unfortunate enough to come into contact with it (but fortunate enough to not die immediately, as many do) into a state of utter delirium filled with terrible hallucinations and partial amnesia that can last for days.

Now the key part of this whole process is that the effects of the tetrodotoxin do not last quite as long as the effects of *Datura*. This means that, when given this mixture (which needed only to be applied to the skin to be effective), the victim first became so paralyzed that they appeared to their family and friends to be dead. Of course, they were not actually dead, and when the tetrodotoxin wore off and they regained the ability to move again, they would find themselves fully in the grip of the *Datura* hallucination, without any sense of who they were, what was real and what was not. Thus, to anyone observing this phenomenon without knowing what was actually going on, the victim would have appeared to have died of unknown causes, and then after a day or so come back to life moaning, talking nonsense, and possibly even being violent. While I'm not aware of any of the Haitian victims of these charlatans having tried to eat anyone's brains (or any other bodily organs, for that matter), it's pretty clear where the idea that these were zombies came from.

Now, obviously, I don't believe that there is any kind of tricky Caribbean sorcerer wandering around the Minecraft world turning innocent villagers into zombies. Rather, I'm providing this history to illustrate how it is possible for certain chemicals, when ingested by a human being, to cause that human to take on many of the characteristics of zombies. After all, if it is possible for such relatively common organic compounds to cause such an effect to humans on Earth, it would certainly seem possible for organic compounds (whether combined intentionally or not) to cause similar effects to the humanlike beings here in the Minecraft world.

There are plenty of other naturally occurring chemical substances, well-known on Earth, that have that same potential to cause this type of alteration in the physiology and brain chemistry of a human. Consider toxoplasmosis, for example. Toxoplasmosis is a disease caused by microscopic parasites that can affect almost any mammal on earth, though the most interesting thing about it is the effect it has on rodents.

Now the first thing you need to know about the parasites that cause toxoplasmosis is that they can only sexually reproduce in the digestive systems of cats (it can be a house cat or lion, but it has to be a feline of some variety), meaning that in order to survive as a species they have to exit the cat in which they were born in its feces, and then somehow make their way into the digestive system of another cat. The problem is that, as you may have observed, cats tend to be relatively solitary animals with rather fastidious personal hygiene (at least for an animal). So, the chance of one cat ingesting the feces of another cat is actually quite small, meaning that these little parasites needed to evolve another method of finding their way into cat bellies for their species to avoid extinction.

Here's where it gets really interesting. This ingenious, evil little organism developed a fascinating strategy for finding new hosts. After the parasites reproduce inside the cat's intestines, their little baby parasites all leave the cat in its feces, where they can survive for many months in a wide variety of environments, including, for example, a small puddle. Now, when a mouse or rat or other rodent comes along and takes a sip out of one of these parasite-filled puddles, they ingest the toxoplasmosis parasites without realizing it. Once these parasite babies find themselves inside a rodent, they head right for its little rodent brain, where they somehow evolved the ability to actually alter its behavior. The ingenious part, however, is how toxoplasmosis alters the behavior of rodents. As you probably know, there are few things that all rodents are more terrified of than cats, and thus they have an instinctual ability to run away pretty darn quickly whenever they see one. However, the way that these particular parasites affect the brain of rodents actually makes them *less afraid of felines*. And what happens to a mouse that isn't afraid of a cat? Well, that mouse isn't going to last too long before it winds up in a cat's belly somewhere . . . which of course is precisely where the toxoplasmosis parasite needs to wind up in order to fulfill its reproductive imperative.

Now, I'm not telling you all this just to gross you out, and I am fully aware that toxoplasmosis is not the same thing as being a zombie (by some estimates nearly half of the humans on Earth have been exposed to toxoplasmosis), but it is an example of how a parasite, capable of transmission from organism to organism, can drastically alter the behavior of the organism it infects. There is even some research to suggest that human beings on Earth who are infected with toxoplasmosis will have their own behavior affected. No, they do not find the need to go around starting fights with house cats, but there is data to suggest that they are more likely to take risks, drive recklessly, and engage in other potentially unsafe or dangerous behaviors at a higher rate than people who are not infected with toxoplasmosis.

I know what you're thinking. Sure, it is theoretically possible for somebody to ingest a chemical or get infected by a parasite that makes them act like a zombie, but that won't make them actually *look like* a zombie. And these creatures here . . . man, do they ever look like zombies. The zombies you find wandering around the forest all look more or less the same: bald, green, scaly skin, no nose, dead black eyes. The zombie villagers, meanwhile, do have the scaly green skin, and their irises have changed from whatever color they were before to red, but they still have their noses, and the whites of their eyes are still, well, white.

Of course, while there are some genetic conditions that can cause a person's skin to change its color in one form or another, these usually involve simply losing pigmentation in the skin (as is the case with vitiligo), or else darken it slightly (as can often be the case with Addison's disease, for example). But these kinds of conditions do not generally cause drastic changes in color like those seen in the villagers. Additionally, except in very rare cases, these types of conditions usually just cause discoloration in patches, or small areas of the skin, and not over the surface of the entire body.

MINECRAFT ZOMBIES

This all got me thinking about the nose. Why did the forest zombies not have noses, while the zombie villagers did have noses? And why would the forest zombies' eyes turn black as well? The only possible explanation I could come up with is that the zombies in the forest were infected with the same condition as the zombie villagers, but they were merely further along in the progression of the illness. Perhaps whatever caused the villagers to change from the way they looked in the village on the other side of the forest to the way they looked in the village on my side of the forest could be a progressive disease of some kind, with its ultimate form looking like the zombies in the forest.

But wait, I thought: *if it is in fact a progression from villager to zombie villager to zombie, with the infected person becoming scalier, greener, and losing parts of its body at each step, wouldn't that make the ultimate form . . . the exploding creeper?*

Yes, I thought, *it probably would*. At least based on the observations I had made, it certainly seemed likely that if I was correct about this being a progressive illness of some sort, the exploding creeper would probably be the progression after the zombie.

Of course, realizing that this was an entirely plausible part of the equation did not make me feel any better. Not just about my own fate should I be unlucky enough to have the same thing happen to me as happened to the villagers, but also in terms of my ability to figure out what the heck was going on. Not only did I have to come up with a theory to explain how some type of transmissible illness could change a humanlike creature's behavior and skin color before causing them to begin losing various appendages, I would also have to come up with a theory as to why that same illness made those same creatures eventually explode.

But what could be capable of doing all that on any possible planet or in any possible universe?

I was thinking about this problem as I walked the short distance from the zombie village back to my cliff dwelling. I got so lost in my thoughts, in fact, that I wasn't looking where I was going and I bumped right into one of the giant mushrooms, badly stubbing my left big toe.

Great, I thought, *now even the fungi are attacking me*.

That's when it hit me: the *fungi*. Of course! How could I not have seen it before? There are only a few species of living organisms on Earth that I am aware of that spontaneously explode as part of their normal life cycle, and all of them are fungi!

Remember back when I was talking about the mushrooms, and how they reproduce by sending out spores. Well, mushrooms are just one kind of fungi obviously, and there are literally millions of different kinds of fungi (at least on Earth) and all of them reproduce with spores. Now, certain types of Earth fungus reproduce with a unique type of spore known as a ballistic spore. If you have ever watched a police show on television you may have seen them talking about taking a bullet to the ballistics lab to see what kind of gun it came out of. This is because the word "ballistic" relates to the science of projectiles like bullets or bombs. The reason why certain types of spores are known as ballistic spores is that they literally explode out of the fungus that produces them in order to find a new place to grow. One example of this type of spore belongs to the very common *Sphaerobolus* genus of fungi.

Often known by their common name of Shotgun Fungus or Artillery Fungus, these neat little fungi grow their spores in fluid-filled sacs. Over time, a chemical process takes place within these sacs that begins to greatly increase the pressure inside of them until eventually, they burst with massive force (one could even call it an explosion) sending the spores flying off to find a new place to create more shotgun fungi. Now, these fluid-filled sacs and the spores they expel are all very small so they aren't really capable of doing too much damage to anything (although if you got them stuck to your car, they are a major pain to clean off). That said, the power of these fungal explosions relative to their size is truly astounding. Some species of artillery fungus have even been measured to accelerate at up to 196,000 meters (643,045 feet) per second squared. That's more than double the acceleration of an *actual shotgun*.

I think by now you see where I'm going with this. Here's my theory:

What if there is a fungus here in the Minecraft world that infects these villagers in such a way as to affect both their skin and their brains, changing their behavior to make them aggressive, while simultaneously changing the color of their skin (or at least covering their skin with the fungus itself)? Furthermore, what if that same fungus, when left untreated long enough on the skin, actually causes the victim's hair and parts of their body to fall off? And what if this same fungus, in order to reproduce, fills its host with spores at increasing rates of pressure until something (like a lost and confused scientist for example) agitates the mentally deranged host just enough for a small rupture to form, causing the entire organism to explode like a bomb, spreading their spores everywhere?

Now, as a scientist, I know that there is no possible way I can say with any degree of certainty that this is what is actually happening. But I can say that it is possible, based on known mechanisms for biological entities on Earth, that something like this *could* happen. And all things considered, the idea of an infectious, parasitic fungus having this kind of an effect on humanoid organisms is still a lot easier for me to understand scientifically than a tree standing up after you remove its trunk.

Lastly, as with so many things here, I lack either the time or the scientific instruments necessary to determine if such a theory is actually true. I did eventually (though not for quite some time after my initial theory of the disease) at least confirm one thing: zombie villagers do in fact start off as regular villagers, and they can be turned back.

I'm skipping ahead quite a bit in time here, but bear with me. Quite a long while after I had more or less accepted my original theory about the zombie fungus, I made about as close a thing to a friend as I probably can here in this world: the butcher from the village on the other side of the forest. You see, while I was perfectly skilled at killing and cooking my own meat, I really, really did not like the killing part of the process. I don't know, maybe it's just having grown up with all the modern conveniences of Earth, but meat is the one thing that I would always trade for, even if I could make it myself. So once every few weeks I would take the journey across the forest to trade emeralds or iron ingots for beef or chicken. And while I never managed to actually have a real conversation with the butcher, I did quite get used to seeing his face, hearing his strange honking sounds, and I'm fairly certain that he looked forward to seeing me as well.

For these reasons, I was more than a little bit upset when, after having my journey across the forests delayed by a particularly nasty rainstorm, I arrived after dark to find the village under attack by even more zombies than usual. I jumped right in with my sword and crossbow (I had gotten quite good at using these by this point) and began helping the villagers to fight off the horde. That's when I saw my friend the butcher in the center of a circle of four moaning, hungry zombies. I ran over to try to help him, but it was too late. Once I had killed all of the zombies, I saw that the butcher had himself been transformed. His skin was green, his irises were red, and he no longer was giving me the friendly expressions and honking noises I had so grown to love.

I tried getting close enough to him to see if there was anything I could do to help, but when he attacked me I fled, terrified that I might catch this awful disease myself. I did not go all the way back to my cliff dwelling, though. Instead, I remained there on the outskirts of the village for several days trying to think of a way to help him. But what could I do? I had no possible way of knowing how this fungus worked, and therefore no way of knowing what could possibly be done to stop it.

Then one night I was crouched on the roof of a house, just watching my friend the zombie butcher moan and drool his way across the village square, wishing I could come up with a way to help him when I saw a creeper slithering toward him. Normally the creepers wouldn't do anything to hurt a zombie villager, but something was strange about this one. He had an odd brownish haze coming off him like I had never seen on another creeper before. It was, however, reminiscent of the strange gas that had once come off my own body after being splashed by a strange bottle filled with nasty-smelling liquid that made me feel so weak that I could barely move my arms. Anyway, for some reason, I just panicked and ran to keep this creeper away from my zombified friend.

Now, the trick to taking out a creeper before it explodes is to go in fast and go in hard. You have to run right up to it and hit it with the most powerful weapon you have before they get a chance to get agitated enough to explode. And that's exactly what I would have done had I not fallen into a ditch just a few meters away from him. As soon as the creeper saw me there, it started making that awful sound that creepers make, and then: *Boom!* I was mostly protected from the blast by the walls of the ditch, but I noticed right away whatever the brown gas was that had been coming off the creeper began to come off of me as well. I saw the brown gas rising off me and just like the previous time, I once again felt like I was so weak that I could barely move a muscle.

Now at that time, whenever I was traveling, I always took with me a whole bunch of golden apples. These wonderful little treats are made by combining eight gold ingots with a single apple. I know it sounds a little wasteful, but they really do taste delicious, in addition to having many health benefits above and beyond those of a normal apple (and besides, even if I found someone to sell the gold to, I would have nothing to spend my millions on anyway!).

I reached in my pocket to grab one of the golden apples, when I saw that my friend the butcher had also been quite close to the creeper when it exploded, as he was also spewing off these strange brown gases. In addition, he did not have the benefit of the ditch to protect him, so he seemed to be quite hurt indeed. I took the golden apple that I was about to eat and threw it to the butcher before hobbling into a nearby house for some shelter. The last thing I saw before I closed the door on the zombie invasion was my friend the butcher, now zombie butcher, spewing not brown but red bubbles and shaking like he was having some sort of a seizure. I fell into the bed, sure that I would wake in the morning to find the butcher either dead or somehow even further transformed.

In fact, the exact opposite of my assumption happened. When I finally saw the first rays of morning sunshine peek through the window, I hopped out of bed and looked outside to see the last remaining zombies bursting into flames and running for the darkness of the forest and the caves. I walked outside to see if I could find whatever was left of my friend, ready for the worst. But lo and behold, when I got to the butcher shop, there he was, right where he had always been, standing in front of his smoker, nonchalantly ready to sell me some meat. Of course, I couldn't actually talk to him about what happened, but he was acting as if nothing had ever happened at all, as if he had never been transformed into a zombie or transformed back.

My assumption was that the cure for the zombie fungus was simply to eat a golden apple, but that did not turn out to be the case. From that point on, I always carried some golden apples in my pocket and always threw them to any zombie villagers I saw, but none of them ever transformed back. My only guess is that whatever those brown gases were that were coming off the creeper, it was in fact them, and not my golden apple (or perhaps even a combination of the two) that finally cured my friend.

In any case, while that is a long way from being any kind of confirmation of my theory about a fungal cause for this world's zombie epidemic, it was the closest I was going to get to supporting evidence of my hypothesis. So for now, I'll just keep carrying my golden apples everywhere I go and avoid eating any mushroom stew!

CHAPTER 9

THE POWER OF REDSTONE

I first became aware of the existence of redstone while attempting to create a clock with a crafting table. This was still back at the beginning of my time here, and I was hoping to find a clock that would work in accordance with the strange, rapid passing of the days that I was experiencing in this universe. Upon finding a recipe for the clock in the crafting table I was immediately struck by two things. First of all, the clock did not really look like a clock at all. If anything, it looked more like a sundial. That wasn't really as strange as the other part, though, since many items in the crafting table looked different once they were actually crafted and existed in full detail. The really strange thing was the ingredients required to make a clock. First, there was gold, which seemed, if nothing else, a little bit excessive. I mean, gold watches have always been quite popular on Earth, but for the crafting table to *require* gold to craft a clock seemed more than a little excessive (not to mention gaudy).

As strange as it seemed to require gold, it was the second ingredient in the clock recipe that really intrigued me: redstone dust. You see, nearly everything listed in the crafting table's recipe book was something that I at least recognized by name from Earth. Even when the item listed in the crafting table was clearly something that did not exist on Earth, I could usually deduce what it was by its name. Furthermore, none of the few items that I did not recognize from Earth (magma cream, blaze powder, slimeball, etc.) had ever yet come up as a necessary ingredient for something else that I wanted to make. So right away I was intrigued. What was this redstone dust? Where could I find it? Why was it necessary for the crafting of a clock? What else required redstone dust to craft?

A quick perusal of the crafting table recipe book let me know that there were in fact quite a few things that required redstone dust, or at least required component items that themselves required redstone dust to craft. Furthermore, it seemed as though the items one could craft with redstone dust comprised the most technologically advanced of all the items in the crafting table recipe book. Some of these technological devices I recognized—the clock, a compass, a piston, a powered rail—but there were also intriguing contraptions whose uses I couldn't even begin to guess, like the mysterious-looking machines called an "observer," a "dispenser," and a "dropper."

One thing was immediately clear: I would have to find some of this redstone dust as soon as possible.

But where did I even look for it? By that point, I had been exploring and collecting samples of every new material I could find for weeks, and not once had I come across any red stones.

My first assumption was that I would have to go on another long expedition—venture out farther than I had yet dared to venture and hopefully find the region of this strange world where all the redstone could be found.

Thankfully, though, before I actually left on my journey, I came up with an even better idea. . . .

I was roaming around the mushroom forest about halfway between my cliff dwelling and the nearby village when I came across a deep canyon. Peering over the edge, I noticed that the rocks at the bottom of the canyon were dotted with all manner of valuable resources: gold, iron, coal, and even copper. Not only that, these blobs of ore seemed to come right to the surface down there, whereas in the past I only found those types of minerals by digging deep mines into the ground. *Who knows?* I thought. *Maybe I'll find some redstone peeking out of the rocks below that I can just scoop up without having to make another epic journey.*

Alas, it did not turn out to be so easy. While there were quite a few interesting and valuable minerals to mine down there, I did not see anything resembling redstone, at least not at the surface. Still, it made me realize that perhaps I did not have to go venturing to faraway lands just to find some redstone. Maybe, on the other hand, I just need to go farther down into the ground to find it. After all, other minerals I had seen seemed to form in larger numbers at greater depths (diamonds, for example) so maybe I just hadn't gone down deep enough yet to find the redstone.

So, instead of gearing up for a long journey, I geared up instead for my deepest mining expedition yet. I emptied my pockets, crafted a few new iron pickaxes and some torches, and made my way down into the canyon so I could start my mine from the deepest point possible.

Then, of course, I dug . . . and I dug . . . and I dug. I dug for twelve days and twelve nights (which, yeah, is really just a few hours), and I found loads of coal, a fair amount of gold, and even a few diamonds . . . but no redstone.

Finally, on the fourth day, my mining tunnel opened up to an enormous subterranean cavern with a huge river of lava flowing down the center of it. I had never found a cavern of this size before, and for a few moments, I just stood there in awe, taking in its massive scale. That's when I saw it: there, right beside the banks of the lava river, was a single grey stone speckled with bright, fire-engine red mineral deposits.

I ran over to the stone and tapped at it with my pickaxe. Amazingly, it began to glow, giving off nearly as much light as one of my torches. Not wanting to waste any more time, I broke it down with my pickaxe and picked up my first bit of redstone dust. There were a few other redstone ore blocks behind that one. And as I began to dig more and more mining caves around the lava river, I found three more redstone blobs, each with four blocks of redstone ore each. So, with my twelve piles of redstone dust safely in hand, I made my way back to my cliff dwelling to start my experiments.

CLOCK

I had collected quite a bit of gold while looking for the redstone, and considering it was the item that inspired my redstone search in the first place, the first thing I crafted with my new material was a clock. I placed a single pile of redstone dust and four gold ingots into the crafting table and waited for it to work its magic.

While I waited, I wondered how much money the gold in this clock would be worth back on Earth and did some quick calculations in my head. I knew that one cubic centimeter of gold weighed about nineteen grams (0.67 ounces), and I knew that it took nine gold ingots to create a one-meter-square (3-foot-square) block of gold. Okay, I thought: *There are 1 million cubic centimeters in a cubic meter. So, dividing that by 9 gets me 111,000 cubic centimeters (4 cubic feet) in each gold ingot. At 19 grams (0.67 ounces) per cubic centimeter, that gave me 2,111,109 grams of gold per ingot, or 8,444,436 grams of gold in my clock.*

Of course, gold is a commodity on Earth so its value is constantly changing, but I seemed to remember seeing that gold was around fifty dollars a gram at one point or another, so I figured I would go with that number. Another few quick calculations in my head and, wow: my clock had $422,221,800 worth of gold in it!

Jeez, I thought to myself as I opened the crafting table to retrieve my finished item. *At over four hundred million dollars (plus however much redstone would be worth), this had better be one heck of an awesome clock!*

And it was . . . well . . . not really that awesome. As it turned out, my initial assumption that the clock looked more like a sundial than a watch face was even more correct than I thought. Basically, the only thing it did was tell you the position of the sun or moon in the sky. From what I could tell, inside the gold casing of the clock was a single flat circle, with one half of the circle colored blue and the other half colored black. In the exact center of the blue half of the circle was a gold hand, and in the center of the black half of the circle was a silver hand. Unlike clocks on Earth, where the hands rotate over a stationary face to show you the time of day, on this clock the entire circle spun around to indicate what time of day it was. At noon, the blue half of the circle took up the entire top half of the clock with the gold hand pointing straight up. From there, the entire circle rotated clockwise, with the black section taking up the entire top half of the clock and the silver hand pointing straight up at midnight.

So basically, yeah, for 400 million dollars I got a device that could tell me exactly the same thing I could learn by looking out the window.

I did, however, eventually find one good use for the clock. It proved quite useful when I was on long mining expeditions underground and couldn't see the sky. Whenever possible, I obviously preferred to emerge from the underground during the daylight so as not to have to battle monsters all the way back to my home, and the clock did make that far easier to accomplish.

The curious thing about the clock, however, was how it worked in the first place. It did not make any sounds at all (or at least none that I could hear) so I knew it couldn't be ticking like a mechanical watch on Earth would. You see, basic mechanical watches and clocks on Earth use variations of a pendulum in order to keep time. A pendulum is simply a weight at the end of a rod or string that swings back and forth in precise intervals of time. If you have ever seen an old-fashioned grandfather clock, the pendulum is the circle hanging from the bar beneath the clock face, inside the cabinet. Now, due to the fact that a pendulum always takes the exact same amount of time to swing back and forth (as determined by the weight of the pendulum and the length of the pendulum arm), it is a fairly simple task to set up gears that will move the hands of the clock a specific distance around the clock face each time the pendulum swings. Of course, a mechanical wristwatch is too small for an actual pendulum (and besides, wearing a watch while walking around would throw off the movement of the pendulum), so they have something called a balance wheel which oscillates back and forth at precise time intervals and accomplishes the same function as the pendulum. This back-and-forth oscillation is what makes the classic tick-tock sound we generally associate with mechanical watches.

And yet, as far as I could tell, this clock did not have any kind of mechanical processes like that taking place at all. Now, on Earth, most modern watches have not used these kinds of mechanical devices since the 1970s, when the quartz-based watch was invented. You see, with a quartz-based watch, a small electric current is passed through a tiny piece of quartz crystal on a circuit board. Now, quartz (at least on Earth) has some rather unique and fascinating properties, the most important in its time-keeping applications being the fact that it is a piezoelectric material. Piezoelectric materials have the unique ability to create an electric charge when mechanical stress is applied to them, and in the reverse, they can actually generate mechanical motion when an electric current is applied to them as well. It is that latter effect that makes them so useful for timekeeping. Not only will quartz generate motion when an electrical charge is applied to it (which is expressed as a high-frequency vibration), the motion it generates will be extremely precise and consistent: when you cut quartz to an exact size and shape, that specific size and shape will always vibrate at the exact same frequency with the application of an electric current. Early quartz watchmakers found that they could cut quartz in such a way as to consistently vibrate at 32,768 vibrations per second. They chose this number (which is 2 to the fifteenth power) as it makes it easy to build a watch circuit that can translate that number to exactly one beat per second. Being that the amount of electricity required to create the piezoelectric effect with the quartz was extremely small, this allowed the production of watches that could be extremely accurate and function for a very long time on a single, small battery.

Well, I thought, *I guess this must be similar to quartz-based clocks then.*
The only problem with this theory was, obviously, that quartz-based clocks require electricity to function and I certainly haven't seen anything resembling electricity here on Planet Minecraft.

Then it hit me: Of course! The redstone itself must be the source of electricity!

Honestly, in retrospect, I can't believe I didn't figure this out sooner. After all, one of the items that the crafting table required redstone dust to make was called a "powered rail." On Earth, the only powered rails I had ever heard of were the kind that ran under electric trains.

Wait, I thought. *Does that mean that I can build an electric train?*
For a moment I let myself get excited about the idea of traveling all over the landscape of Planet Minecraft on a network of fancy, comfortable electric trains until I remembered how long it took me to collect a measly dozen piles of redstone dust. Collecting enough to cover the landscape with powered rails would probably take me an entire lifetime. . . .

Still, the prospect of the new avenue for my scientific exploration was just as exciting. Somehow, in some way, this redstone dust seemed to be the source of electricity here in this world, and I was going to find out how it worked and what it could do.

COMPASS

Okay, before I get into some of the more fanciful electronic-type devices one can build with redstone, I have to talk about the second thing I made: the compass. And yeah, it's a curious one.

I guess I should have known better by this point than to think that a Planet Minecraft compass would work like an Earth compass, but in this case, the representation pictured in the crafting table really did look exactly like every other compass I had ever seen back home. It was a metallic cylinder, hollow and open at the top with a red dial pointing from the center to the edge. No, it didn't have any directional markings like north, south, east, or west on it, but it was clearly a compass.

The recipe for the compass was exactly like the recipe for the clock, just substituting iron for gold. Of course, I did wonder to myself, while the crafting table was working, how much money's worth of iron would be going into this compared to the exorbitant amount of money worth of gold that went into my clock. I recalled reading once in an article that the price of iron was about the same per metric ton as gold was per gram, which should make this a pretty easy conversion. There are 1 million grams in a metric ton, so if the clock used 422 million dollars' worth of gold, then my compass was using $422 worth of iron. This sounded pretty good, but then I realized my mistake: gold is about two and a half times as dense as iron, which means that each of my iron ingots would weigh about two and half times less than my gold ingot. So, doing some more quick math, I figured out that the actual cost of the compass in iron would be a measly $168.80. What a bargain!

Now on Earth, a compass is simply a magnetized piece of metal suspended at the middle of a dial so that it may freely spin around. Why does it always point north? Well, that's pretty simple, actually. You see, the Earth itself is kind of like an enormous magnet, and like any magnet, it has both the north and the south pole. The thin lightweight piece of magnetized metal also has a north and a south pole and, if you recall from when we talked about magnetism earlier, opposites always attract. Thus, the south pole of the compass needle always points toward the magnetic north pole of Earth. Simple.

At first, it seemed like this Minecraft compass was more or less going to work the same way an Earth compass would. While it didn't point to what I had been thinking of as north, I didn't actually know which direction north was on this planet. From my vantage point, the sun rose on one side of the planet and set on the other side of the planet so I called those two directions east and west, respectively, assigning their perpendicular neighbors as north and south.

From where I was standing in my cliff dwelling, the compass I had just crafted seemed to be pointing almost exactly due west (which is to say, in the direction of the setting sun). This alone did not mean that my new compass worked any differently than any compass on Earth did. If anything, it probably just meant that the magnetic pole on this particular planet was somewhere far off to the west of me. As far as I was concerned, though, this was just as good as a compass that faced north. As long as I knew which direction was which, I could use it to help me get around when I couldn't easily see the sun or the moon.

Over the next few days I mostly just stayed within a few hundred meters of my home, and every time I pulled my new compass out of my pocket it just kept pointing west. A little while later, however, when I was venturing quite far to the north, I noticed that the compass seemed to be pointing a little bit to the south of the setting sun. Then, a little while later, while I was venturing far to the south, I noticed that the compass was pointing a bit to the north of the setting sun. I didn't think too much of these variations when I saw them, though, as I assumed they were just the result of some nearby magnetic mineral deposits or something.

It wasn't until quite some time later—maybe a few weeks at least— that I first noticed something really strange about the behavior of my compass. I had been exploring a region in the far north when I decided to change direction and travel west for a while. Now, my assumption was that I could simply follow my compass due west, and it would lead me more or less in a straight line, and then when I wanted to change and go back, I could just turn around and keep the compass needle pointed directly behind me. But as I began to move west in a straight line, I noticed that the direction of my compass needle began to shift. At first it was just a small shift, but as I kept traveling west, it clearly kept shifting farther and farther to the south. It was raining at the time and I couldn't see the sun, so my first guess was that I was simply letting myself turn slightly to the north without realizing it. So, I did what I figured was the right thing to do, and I turned to the left until the needle was once again facing straight up on my compass and began to walk straight again, assuming I was once again traveling due west. I walked this way for quite some time, until, to my great surprise, the clouds finally cleared away and revealed a setting sun that was not directly in front of me, but rather about 45 degrees off to my right. Not only that, my compass was no longer pointing anywhere close to the direction of the setting sun.

Well, I thought, *this could only mean one of two things: either the compass is broken, or else the magnetic pole on this planet must be a lot closer to me than I realized.* Normally, on Earth, we don't pay much attention to this, as most of us don't spend too much time wandering around the magnetic north of our planet (which is not actually at the North Pole, but rather somewhere in the northern reaches of Canada) but, if you were to fly a plane just a little bit east of due north on Earth until you began to get close to the magnetic north pole, the compass would do exactly what my compass is doing now: it would start to turn to the left even while you continued to fly in a straight line. Thus, if you followed the direction that the compass needle was pointing, it would lead directly toward the magnetic north pole. So, logically, I must be close enough to the magnetic north (or magnetic west, as it were) of this planet to have made a partial semicircle around it. This, of course, would mean that if I were to follow the compass directly from where I was standing for long enough, I would eventually wind up right on top of that very magnetic pole. Furthermore, it should be super obvious when I have arrived at the pole, because the second I walked past it, the compass would spin around and point in the opposite direction.

I checked my supplies to confirm that they were still quite plentiful, and decided that this was too good an opportunity to pass up. I turned fully in the direction that the compass needle was pointing and marched on in search of the magnetic pole of Planet Minecraft.

The farther I traveled, the more I began to realize that my surroundings were looking increasingly familiar. I probably would have recognized the exact place I was being led even sooner had I not been approaching from an entirely new direction than I had ever approached before, but by the time I finally made it to the spot where the compass flipped, there was no doubt at all about where I was and it sent shivers down my spine. I was at the exact spot—not the same region, not nearby, not even a few yards away—the *exact* same spot where I had first appeared in this world after leaving my own.

Once again, there were two possibilities as far as I could tell: either I had arrived on this planet at the exact location of the northern magnetic pole, or else something about my arrival here on this planet created a magnetic field at that exact location that was powerful enough to interfere with my compass's ability to point toward the actual magnetic north pole.

I decided upon reflection that, from a practical standpoint, it really didn't matter which theory was the correct one. As long as I knew where my compass was pointing (and it always pointed at the same place), I could use it to orient myself just as well as if it always pointed north. Besides, figuring out which of those two theories was right would now require me to figure out how I wound up on this planet in the first place, and I certainly wasn't ready to do that. At least not yet.

REDSTONE LIGHTS

Okay, enough with compasses and clocks, and on to the really interesting stuff: redstone electricity. Even before I ventured out to learn the truth of my compass, I was eager to see if my theory about redstone being a source of electricity had any merit to it or not. The first thing I crafted in order to figure this out was a redstone torch. I was hoping that this torch would wind up simply being an electric light on a stick that would instantly prove my theory correct, but alas it looked more or less like a regular torch, save for the color of the flame, which was, as you may guess, decidedly red. Just like a regular torch, it came out of the crafting table about the size of a matchstick and then lit itself and expanded in size as soon as I placed it on a surface. While it wasn't quite as bright as the standard torch and produced a bit less smoke as well, I really couldn't tell any difference. At the time, I assumed that all I was going to learn about redstone from the torch was that this strange mineral was flammable in a fashion similar to coal, regardless of its other attributes.

This assumption turned out to be totally wrong, but I didn't figure that out until after I had crafted the next item on my list of available redstone recipes: a redstone lamp.

The recipe for a redstone lamp requires four piles of redstone dust and one glowstone. Thankfully, I still had some glowstone left over from my trip to the Nether (I was in no rush to go back there at that point in time) so I loaded up the crafting table with the required ingredients and waited to see what popped out. Honestly, I did not have terribly high hopes for learning much from the lamp, if only because the glowstone itself emitted a fair amount of light, so I figured I probably wouldn't see much more with the lamp than a regular piece of glowstone prettied up a bit and colored red.

As soon as I took the redstone lamp out of the crafting table, though, I knew something was different right away. Both the regular torch and the redstone torch that I had crafted had lit up the moment I placed them on a surface, which was how I expected the redstone lamp to work as well. When I placed the lamp down on the ground, however, nothing happened. Even with the glowstone in there, my new lamp wasn't putting out any light whatsoever; not a single, solitary lumen.

I ran my hands over the entire surface of the lamp multiple times in search of a hidden button, but I couldn't find anything. I shook it around, placed it in different locations, but nothing I could do would make it light up. I even tried putting a redstone torch directly on top of the thing and still nothing. Finally, after I'd given up on everything else, I decided that I still wanted some form of light there (I was trying to set it up right next to my bed so I could have a bit of a night light when I was sleeping), so I gave up on the lamp and just placed the redstone torch down right beside my bed so I could relax in its warm glow while I got some rest for the evening. And wouldn't you know: the second I placed that redstone torch down on the block beside the lamp, the lamp lit right up.

What the heck? I thought. *How is that even possible?*

I picked up the redstone torch and the light went out. I put it down again on a different side of the lamp and it lit right up again.

Okay, I thought. *Clearly, the torch is providing power to the lamp. Now I just need to figure out how it works.*

The first thing I did was to move the redstone torch one block farther from the lamp to see if it would still power it up. It did not. Next, I tried moving the torch another block farther away and, unsurprisingly, it still did not power the lamp. I already knew that placing the torch on top of the lamp did nothing, but what about on its side? It tried this: still nothing. The next logical step was to try placing the torch beneath the lamp so I dug out the space beneath the redstone lamp (which, like so many other things here, remained suspended where I had originally set it regardless of the fact that there was nothing underneath it) and placed the redstone torch on top of the block beneath it. Finally, the lamp lit up again.

I then tried placing the redstone torch two blocks beneath the lamp with an empty block between them, and while that did nothing, when I placed a block of clay between the torch and the lamp, it lit right up just as bright as before.

Now we're getting somewhere, I thought.

Clearly, the redstone torch was producing some form of an electric-like field around it that could only travel so far through the air but could travel more easily through solid materials. This, as it turns out, is quite similar to the way that electricity works on Earth. If you have ever seen two wires create a spark between them, you will know what I'm talking about. Let's say you have a simple copper wire that is two meters long, with an electric current running through it from a power source to a light bulb. When you turn on the power source, the wire will carry the electric current to the light bulb and the light bulb will illuminate. If you cut that wire right down the middle and move the two cut ends a meter apart, what will happen? Simple: the light will turn off. However, if you move those two raw ends of wire very close to each other, but not quite touching, what will happen then? A spark (known as an arc, in this case) is what will happen then.

An electric arc, also known as an arc discharge, is that spark of blue light that travels through the air from the charged side of the wire to the other side. If there is enough current running from the battery, the light bulb will still light up, even though there is a small gap in the wire itself. The blue light you see in the gap, or the arc, is simply the air molecules being heated by the electricity passing through them as it travels from one end of the cut wire to the other. This heated, electrically charged air is called plasma.

Now, if you move that wire just a little too far apart, the current will not make the jump across, and the light will not illuminate. However, if you were to place a conductive material (like, for example, a copper cube) at the right space between those two wires, the current can jump from the wire to the copper cube and then from the cube to the other wire, and from the other wire to the light bulb.

This, I reasoned, must be similar to what is happening here. The redstone torch emits some kind of an electric current that can only travel so far through the air. When you place the redstone torch directly beneath the red stone lantern, it is close enough for the current to reach the lamp and light it up. Moving the torch an extra block farther down makes it too far away for the current to travel through the air into the lamp, so the lamp remains dark. Placing a block of conductive material between the two allows the current to jump from the torch into the conductive material and from the material to the lamp, which then illuminates. Of course, the block of clay that I used would certainly not be conductive on Earth (in fact, ceramic is often used as an insulator on electrical power lines specifically because it does not conduct electricity) but that's a matter for another time. . . .

The only real complication I ran into with this theory at the outset was that it does not seem to work when the lamp is placed two blocks to the side of the torch with that same conductive material between them. My guess about this at the time, however, was that it had to do with the fact that the torch head, which is clearly where this current is coming from, sits at the top center of the square meter block that it takes up. This means that it is quite close to the conductive material placed above it but a little bit farther from the conductive material placed to the side of it. The current seems to be enough to reach the lantern when the torch itself is placed directly beside the lantern, but the extra juice it would need to make the jump through a conductive material and into the lantern surpasses the overall charge of the torch by just enough for it not to work.

Over the next few weeks, I spent a great deal of time experimenting with redstone torches and lamps. I tried using different materials as conductors, different arrangements of lanterns, etc. All in all, I came away from these experiments with a lot of good information. I was slowed down a little bit by the fact that I wound up having to make another trip to the Nether for some glowstone, but thankfully there was a nice big batch of it close to the portal, so I was able to run in and out without even having to get into a fight with any of those horrible Nether creatures.

One of my key findings from all this research was to determine which materials worked as a conductive middle layer and which ones did not. Some of the materials that conducted redstone power were not that surprising; substances like metal, stone, bricks, and wood all had no trouble conducting the energy, while ice or glass, for example, did not. The conductive properties of other materials, on the other hand, were far less intuitive. For example, a full block of wool had no trouble whatsoever conducting power, but an actual block of glowstone (which, remember, is the only ingredient in a lantern other than redstone dust) could not conduct the power at all.

It wasn't until I had a little bit of an incident involving leaving a door open at night and some wandering zombies making their way into my house, that I made the discovery that really changed everything.

I had been losing track of time quite a bit during my research, which was rather easy to do when the days and nights only last twenty minutes, and this particular night was no exception. It was getting close to midnight, and I had been working on my redstone research for days when I was standing at the crafting table and heard the familiar, terrifying moan of a creeper right there in my cliff dwelling. I turned around and saw the terrifying creeper, wide-eyed and ready to blow himself up right there in the middle of my home. Knowing that letting him explode in the middle of everything could destroy weeks of work, I sprang into action, took out my sword, and tried to slay the beast before he could blow.

Alas, I wasn't quick enough. Thankfully, I did manage to keep him away from the part of my house where I conduct my experiments, so the only thing he blew up was a section of the exterior wall, my bed, and a chest filled with various items that got scattered all over the floor. I quickly set about rebuilding that small section of the floor and making sure all of my doors were closed, before going about repairing the wall. I was trying to move so quickly though, and I was so very tired by that point, that I didn't notice that instead of laying down red terracotta blocks, I laid down a few meters of redstone dust in a line on the floor—a line that just happened to run right between where I had an unlit redstone lamp sitting on the ground and a redstone torch a few meters away.

Instantly, the lamp lit up. Not only that, the line of redstone dust itself began to glow as well.

It's a wire! I thought, as delighted as I was amazed. Laying down redstone dust in a line creates a wire capable of transmitting the power from a torch to a lamp.

This, of course, was a monumental discovery. Not only did it mean that I could power a lamp without having to place a redstone torch so close to it, but the ability to transmit power in this fashion paved the way for literally dozens of other useful devices. I could easily fill another five hundred pages just describing all the things you can create with redstone, but in the interest of sticking to my mission here, I'm instead just going to take a few minutes to discuss the truly unique properties of redstone, and my theory for how and why it functions the way it does.

Now, you have certainly gathered by this point that redstone is, in and of itself, a source of energy. Not only that, the energy it produces is quite similar to (though not quite exactly the same as) electrical energy on Earth. This alone is quite extraordinary, as there is no naturally occurring mineral on Earth capable of producing a stable electric current all by itself. Sure, we have radioactive minerals that produce radiation that can be used to generate electricity, but it's not as if you can just plug a light bulb into a block of uranium. Of course, you do have to do a little crafting to your redstone dust to turn it into a power source, but you don't actually have to add anything else to it for it to generate power. Simply crafting a one-meter-square (3-foot-square) block of redstone out of nine piles of redstone dust gives you a completely usable power source (and one that you could build a house out of, if you wanted to).

Perhaps the most unique and mysterious quality of redstone power, however, is its longevity. Even the simple redstone lamp, powered by a single redstone torch, that I placed beside my bed back when I first started these experiments is still lit up as bright as when I first turned it on. Of course, I don't actually think that redstone power is infinite, as that would seem to go against even the wacky laws of Planet Minecraft's physics, but it is clearly far more efficient and long-lasting than any source of energy we have on Earth.

This brings us to the big question about redstone: How does it create energy so efficiently?

I pondered this question for quite a long time, and ironically, it turned out to be the first thing I crafted with redstone that led me to finally come up with a theory to answer this question. That's right: the 400-million-dollar clock.

If you recall, my eventual theory about how the clock kept time revolved around it being somehow similar to a quartz watch on Earth, using the piezoelectric effect to keep time with some kind of unknown internal mechanism. Something about this theory always bothered me though: Where was the quartz? Quartz does exist here on Planet Minecraft (well, technically the quartz itself comes from the Nether, but still, you can find it here), and yet the recipe for the clock didn't actually require quartz at all.

I was puzzling over these questions in my head one day while mining for some redstone ore deep beneath the ground when I slipped and bumped my pickaxe into a block of redstone ore that I had yet to mine. As it did the first time I found redstone ore, the block began to glow. By this point, I had seen this effect so many times that it was hardly noteworthy, but the fact that I was just thinking about the clock made it all click for the first time: The redstone began to glow *when I hit it*. In other words, it began creating energy by the application of mechanical force—If that wasn't a piezoelectric effect, I don't know what is. Not only that, but the glowing continued for quite some time—far longer, in fact, than could be attributed to the mechanical energy I added to the redstone ore block by hitting it.

This redstone clock didn't need quartz to create a piezoelectric effect, because the redstone itself was a piezoelectric material!

Here's where it really gets interesting. You see, due to the fact that, as mentioned above, the piezoelectric effect is reversible, it is possible to apply mechanical stress to a piezoelectric material, then take the electricity it generates and route it back to the material to generate mechanical motion, which you then route back to the material to generate electricity. Of course, a certain amount of energy is lost every time you reroute it back to the material, so this effect will not last forever (though scientists on Earth have made some very efficient self-charging piezoelectric batteries) and if you try to get this to power anything else (like a clock, for example) the energy will diminish very quickly.

If, however, your piezoelectric material also contained a large amount of easily accessible potential chemical energy (like the energy stored in petroleum, natural gas, or coal, for example) you could theoretically create a piezoelectric circuit that accessed this stored potential energy to create a self-charging piezoelectric battery that would last an extraordinarily long time.

Of course, I'm not suggesting that the crafting table is actually creating any kind of super-complex circuitry inside every redstone torch or anything. My theory, rather, is this: redstone, in its solid form (not as a dust), is a piezoelectric material that, at the molecular level, is structured in such a way as to create a piezoelectric loop that automatically feeds and reflects its energy back on itself (meaning that the mechanical energy generates electrical energy, which in turn generates mechanical energy, etc.) while simultaneously consuming its own stored potential energy. The result of this is an extremely efficient energy source that needs only to be activated (either by the application of a small amount of mechanical or electric energy) and then continuously generates electricity until it consumes every last molecule of itself (which, given the material's incredibly efficiency, would take an extraordinarily long time).

I know, I know . . . the actual scientists reading this will no doubt take issue with my complete and utter lack of evidence, and probably have a million reasons why this wouldn't work on Earth. To them, I say this: I am not on Earth. I am on a planet with a square sun, impossible gravity, and spiders the size of a Harley Davidson. If you want to prove me wrong, figure out how to bring an entire university physics lab through the Minecraft portal and have at it. Until then, I'm sticking with my theory.

CHAPTER 10

THE END

Well now, this chapter is called The End, and though it is the last chapter (at least in this volume) about my research and explorations here on Planet Minecraft, it is certainly not the end of my journey on this world.

That being said, there is no place better to start than at the beginning of the end. Not long ago (though long after I had lost track of how many days and nights I had spent in this world), I was surprised on a dark hillside by a tall, black, ghostlike creature that looked as if it would be far more at home in the Nether than the Overworld. I had seen this type of creature a few times before, though I had always been quite satisfied to leave them alone, as they had never before appeared to pose any kind of a threat to me. On this night, however, things did not stay so peaceful. Perhaps it was just the shock of having this ghostly being appear out of nowhere right beside me when I was already looking over my shoulder, but before I could stop to think I just lashed out with my sword and struck the creature down.

Like every other living thing on this planet, the creature disappeared in a puff of smoke, but what it left behind was something that I had never seen before. The object looked like a large blue marble, faintly glowing and seeming to give off an almost imperceptible vibration. Always excited at finding a new item to research, I ran straight back home and popped it into the crafting table to see what it was called and if it could be used to make anything interesting.

According to the crafting table, the object was called an Ender Pearl and it was only useful in the crafting of one item. By adding it to some blaze powder (which I had long ago brought back with me from the Nether), the crafting table would transform this strange gem into something called an Eye of Ender.

Well, that's intriguing, I thought to myself.

I went to my chest, rummaged around a bit, and found the blaze powder. I popped it into the crafting table beside the Ender Pearl, closed the lid, and a moment later I was holding an Eye of Ender in my hand. It didn't really look that much different than the Ender Pearl, save for its greenish hue and a decidedly feline pupil shape at the center.

Okay, I thought. *Now, what the heck do I do with it?*

My first idea was that it would probably be something that I could use with one of my redstone devices, as most of the rare materials I crafted seemed to have something to do with redstone. So I took it out to my workshop and tried to place it next to a redstone lamp to see if it would have any effect. As soon as I tried to set it down, however, the Eye of Ender flew right out of my hand, climbed about ten meters (33 feet) into the air, hovered for a second, and then fell to the ground ten meters (33 feet) away from where I was standing.

I walked over, picked it up, and looked at it to see if anything had changed about the object, but it looked exactly the same. Assuming whatever its purpose was had to do with its short flight I tossed it into the air, trying to make it fly back to the other side of my workshop. To my surprise, though, it didn't go in the direction that I threw it at all. Instead, the Eye turned around and floated off behind me. Once again it went up to ten meters (33 feet) up in the air and landed ten meters (33 feet) away in the exact same direction that it had come before.

I tried throwing it in every possible direction, but it always wanted to go the same way. The only other thing that I had crafted that behaved remotely like this was my compass. Now, considering that my compass always points back to the place where I began my journey in this world, I couldn't help but wonder if this Eye of Ender would point me in the direction of where I would end my journey. Not knowing how far this strange object would take me, I picked up some bread, some torches, weapons, and pickaxes, and followed the Pearl to see where it would lead.

It soon became obvious that the Eye of Ender was pointing me in a very familiar direction. From my cliff dwelling, it went in a straight line directly toward the zombie village I had found so long ago (which, since I had finally gotten around to curing all the zombie villagers by this point, had become just a regular village). I assumed it was a coincidence though, and that it would leave me somewhere far beyond the village, but quite the opposite turned out to be true.

Before long, the Eye of Ender had led me all the way to the road at the edge of the village. Not wanting to freak out any of the villagers (just in case this was some kind of dangerous, mystical talisman or something) I waited until nobody seemed to be looking in my direction before I tossed it in the air again. By this point, I had just passed the fountain at the center of town, and to my great surprise, the Eye flew not in the same northerly direction it had been going up until that point, but instead took a hard right, flew a far shorter distance, and then dropped right beside the fountain. I picked it up, walked a little farther away, and tried to throw it again to the north, and again it flew behind me and dropped right next to the fountain. No matter where I went, or which direction I threw the Eye, it always landed right there beside the fountain.

I deduced that my initial assumption that this was some kind of a compass-like directional device was probably correct, but I had no idea what it was trying to lead me to. There didn't seem to be anything particularly interesting about the fountain itself, so I figured whatever the Eye of Ender was trying to reach must be under the fountain. After all, I had never actually tried mining underneath the village itself, as it had never really seemed necessary before. But what else was I going to do? I had to see if there was anything down there.

Now I didn't want to start digging a mine shaft right in the center of town, so I moved a little ways off, just past the edge of the village, and began to dig. I dug straight down for about ten meters (33 feet) and then began to dig back in the direction of the fountain.

After stopping to mine out a little vein of copper I found (I never miss an opportunity to pick up some choice metals), I made it to the spot directly beneath the fountain. Once there I just began to dig all around me to see if I could find anything interesting. Maybe there would be some huge cache of redstone bigger than anything I'd seen before, I hoped. Of course, I also realized that it was just as likely that there would be some kind of a horrible monster I had never seen before, but that's why I brought my sword.

I dug out a large, 10-cubic-meter cavern that went right up to just below the fountain, but I didn't find anything more interesting than some copper and some coal. So I dug some more, increasing the size of my cavern to 20 cubic meters (706 cubic feet), but I still found nothing.

So I took out the Eye of Ender again, hoping that if I tossed it down there beneath the ground it would show me something new. I let it fly, and this time it just flew straight down and disappeared into the rock beneath my feet.

Not wanting to lose this fascinating new treasure, I dug straight in the direction that it went without even trying to widen my cave, hoping that it would be waiting for me somewhere below. I dug through granite and cobblestone and even some gravel, but I still didn't find it. Or rather, it wasn't the first thing I found. The first thing I found was quite surprising: a single block of mossy stone bricks right there beneath a block of granite that I had just dug up.

Realizing that this must be some kind of a subterranean structure or mine shaft that the Eye of Ender was leading me to, I quickly dug into it and found myself inside what appeared to be some kind of underground fortress. To my great relief, there in the hallway beside a small set of stairs was the Eye of Ender. I picked it up and began to explore this strange subterranean structure.

To say that this underground fortress was a labyrinth would be an understatement of enormous proportions. It has to be one of the most mind-boggling confusing places I have ever explored on Planet Minecraft. I was almost instantly lost, and while I did find a few chests with some interesting items in them, mostly what I found down there were zombies and silverfish. Also, it became very clear that whatever was attracting the Eye of Ender was only meant to bring its bearer into the fortress, and not to any particular point inside of it.

No matter where I tried to toss it, the Eye always just flew right back to that spot beside the stairs. Still, I had to believe that any item so rare and difficult to come by that I actually had to kill one of those strange black creatures in order to get had to be leading to more than just an abandoned warehouse under a zombie village. It was also quite clear, however, that I was not going to find it easily.

Over the course of the next few weeks, I frequently returned to explore more of the underground Fortress, always managing to discover a new hidden room, stairway, or chest filled with interesting items. It took about a month of doing this to finally find the room that the Eye of Ender had clearly been trying to lead me to. After getting myself very turned around and lost in one of the many identical hallways of the fortress, I grew frustrated and decided to simply cut a hole straight back in the direction that I thought my staircase to the surface was. Along the way, I unleashed a veritable torrent of silverfish, which I barely managed to fight off. When I finally looked up, I was in a room different than any I had seen there before. It was bigger than most of the rooms in the fortress, with two lava pits on either side of the entrance, and a staircase right in the middle of the room.

At the top of the staircase was a metal cage that silverfish seemed to be spawning out of almost constantly. I quickly ran up the stairs and smashed the silverfish spawner only to find myself staring at something quite unique indeed. It was a square of golden-green boxes floating over a pit of lava. There were three boxes on each side of the square, for a total of twelve.

At the top of eleven of the twelve boxes was an empty slot that looked like it was missing some vital piece. Thankfully, I only had to look at the twelfth box to realize exactly what they were missing because in that box was another Eye of Ender, just like the one in my pocket. I pulled the Eye out and placed it in the slot beside the other one. It made a satisfying chime as it dropped in, fitting perfectly. Then, I waited. But nothing happened. Clearly, I was going to have to find myself another ten Eyes of Ender.

In the interest of brevity, I won't tell you the whole boring story about how long it took me to actually acquire ten more Eyes of Ender, or how bad I felt about having to kill all those Endermen to get them, but I just had to know what would happen if I filled all of the boxes.

Finally, after a few more weeks, I had everything I needed and went back to the Fortress to test it out. I climbed to the top of the staircase, placed the rest of the Eyes in the remaining empty slots, and then: *WHOOSH!* The empty space in the middle of the square filled with blackness, and then the blackness filled with glowing, floating particles of light.

Clearly, this was another portal. But where did it lead? Could my initial guess have actually been correct? Could this be the end of my journey? The way home?

I had to find out. I held my breath, clutched my sword, and dove in.

I was expecting . . . well, I don't really know what I was expecting. I think I was expecting to either wind up right back in that laboratory in Mojang Institute or else to find something like what I found on the other side of the Nether Portal: fiery lava lakes, scary monsters, flowering vines, mushrooms, zombie pigmen, and everything else that I found back in that hellish dimension.

As it turned out, both of my guesses were wrong. It was nothing like the Nether, and sure as heck wasn't the Mojang Institute. It was just a room, and a small room at that: five meters (16 feet) wide, five meters (16 feet) long, and three meters (10 feet) tall. The floor appeared to be made of obsidian, while the walls were made out of some strange cheese-like green substance that I had never seen before—not on Earth, and not on Planet Minecraft either.

It didn't smell like anything, and it didn't even feel like anything. It wasn't even hot or cold. I know it's impossible, but I would almost describe it as not having any temperature at all. One thing was clear, though: there was no portal in that room, and there was no door, which meant that the only way to get out of there would be to dig.

I knew from experience how hard obsidian was, so I figured that whatever the walls and ceiling were made out of would have to be softer than that. Not wanting to wear out my pickaxe any sooner than was absolutely necessary, I decided not to even try mining the floor so I went to work digging through the walls. Thankfully, the first block gave way with a single strike from my pickaxe.

I figured that I would just dig in one straight line and if I didn't find anything in a hundred meters (328 feet) or so, I would turn around and try digging the other direction, and just keep going like that until I found something.

As it turned out, it was barely 30 meters (98 feet) before I struck through the last block to find a vast black expanse stretching out in front of me.

I emerged from my tunnel on the side of what seemed to be a large island floating in an absolute void. Turning to my right I saw what appeared to be a nice easy slope up to the top of the island, so I made my way over and began to climb. As soon as I reached the top, however, I wished I had never come through that portal at all.

The entire top of the island was one vast plain, absolutely covered in Endermen, and all of them were staring at me. As if that wasn't bad enough, when I looked beyond the sea of ghostly black creatures, I saw, right at the center of the plain, a set of huge obsidian towers rising high into the sky, and flying in the middle of the towers was an enormous dragon.

Unfortunately, the second after I saw the dragon, the dragon saw me.

Thinking that maybe this really would be the end of my journey, after all, I raised my sword, and decided right there and then that if I was going to go down, I would go down fighting.

I never even got the chance to swing my sword. Within the span of a heartbeat, the dragon was flying down on top of me as the Endermen swarmed. I felt a rush of pain as the Endermen began to strike, and then the dragon swooped down with its gigantic claws, and then—

Then I was right back where I started. Literally. I was right there, in the exact same spot where I had first opened my eyes on Planet Minecraft. I don't know how, and I don't know why, but that's where I wound up. Eventually, I'm sure I will come up with a theory for all that, too. But not yet. For now, I'm just going to accept that some things are beyond my explanation, and when I finish this book I'm going to leave it right there in that same spot too.

Who knows, maybe someday Dr. Charles Benzak will make his way back here and find it.

If that's you, Charlie, come say hello. Just follow the square sun in the morning and hang a right when you get to the giant mushroom forest. My house is the big one at the top of the red mesa.

And don't worry: I'll leave the light on for you.

INDEX

Copyright © 2022 by James Daley

Skyhorse Publishing books may be purchased in bulk at special discounts for sales promotion, corporate gifts, fund-raising, or educational purposes. Special editions can also be created to specifications. For details, contact the Special Sales Department, Skyhorse Publishing, 307 West 36th Street, 11th Floor, New York, NY 10018 or info@skyhorsepublishing.com.

Skyhorse® and Skyhorse Publishing® are registered trademarks of Skyhorse Publishing, Inc.®, a Delaware corporation.

Visit our website at www.skyhorsepublishing.com.

10 9 8 7 6 5 4 3 2 1

Library of Congress Cataloging-in-Publication Data is available on file.

Cover design by Brian Peterson
Cover image by James Daley

Print ISBN: 978-1-5107-6775-1
Ebook ISBN: 978-1-5107-6776-8

Printed in the United States of America